T0191684

Changes in Production Efficiency in China

Bing Xu · Juying Zeng
Junzo Watada

Changes in Production Efficiency in China

Identification and Measuring

Springer

Bing Xu
Research Institute of Quantitative Economics
Zhejiang Gongshang University
Hangzhou
People's Republic of China

Juying Zeng
Institute of Quantitative & Technical Economics
Chinese Academy of Social Science
Beijing
People's Republic of China

Junzo Watada
Graduate School of Information, Production and Systems
Waseda University
Kitakyushu
Japan

ISBN 978-1-4939-5388-2 ISBN 978-1-4614-7720-4 (eBook)
DOI 10.1007/978-1-4614-7720-4
Springer New York Heidelberg Dordrecht London

Softcover reprint of the hardcover 1st edition 2014

Printed on acid-free paper

Springer is part of Springer Science+Business Media (www.springer.com)

Acknowledgments

This research was supported by the project of National Natural Science Foundation of P.R. China: "Identification and Simulation on Change of Production Efficiency with Investment Path-Converged Design (No. 70973110);"

the Fifth Chinese National Postdoctoral Special Sustentation Fund Project (2012T50172); and the Major Project of Science Foundation of Zhejiang Province (No. Z6110519).

Contents

Abstract

Against the background of sustainable growth, the Chinese government has put forward both stimulating investment and accelerating healthy urbanization development as priority tasks in following 5 years in 2013. The initial investment and resource allocation philosophy in China is oriented with equity allocation, while the investment philosophy of foreign direct investment (FDI) emphasizes efficiency promotion. The absorbing of FDI will inevitably bring about changes in production efficiency with the transfer from equity pursuit to efficiency promotion. Therefore, the fast rise of the proportion of fixed asset investment to GDP and population agglomeration has naturally raised questions as follows. First, whether the investment and population agglomeration brought changes in production efficiency in the past China? Second, whether the promoting role of investment, especially foreign direct investment (FDI), on economic growth in Eastern China can be transferred to the growth in both Middle and Western China in future?

Synthesizing the perspectives of management science and statistics, the book aims to identify and measure how investment and population agglomeration changes in production efficiency in sustainable growth from aspects of allocation efficiency, scale efficiency, technical efficiency, and sustainable efficiency with innovative path-converged design techniques. The main results are presented as below.

Theoretically, using large sample theory from statistics achieves unique results and the ability to reproduce experiment in engineering; the technique of path converged design is innovative in two aspects. First, path identification presents an observable benchmark as the criterion of the production efficiency to replace the unobservable production frontier surface. Second, path-converged design is designed to select a controllable variable as a path of identification and to ignore uncontrollable natural variables.

- *Allocation efficiency with investment and population agglomeration*
  Reallocation of public expenditures in cities successfully reduces allocation inefficiency by up to 2 %, with reallocating 27.83 billion Yuan of public expenditure, which accounts for 2.020 % of the original total amount in overall China in 2006 (Chap. 2).

Reallocation shows that decision-making eliminates allocation inefficiency in urbanization development with population migration from lower urbanization levels to higher urbanization levels. The inefficiency of regional population patterns shrink to 5.800, 4.100, and 5.600 % for small, medium, and large cities, respectively (Chap. 3).

- *Scale efficiency with investment*
  Foreign direct investment (FDI) path exhibits a crowd-in effect on capital after 1996 in the Eastern region, while crowd-out effect on capital is presented in both Middle and Western regions during almost the entire period. Possible strategies are designed with environment FDI path identification to narrow regional discrepancy. Implementation of urbanization environments with the FDI path realizes the crowd-in effect on capital in Middle regions. However, the situation fails in Western regions in China (Chap. 4).
- *Technical efficiency with investment*
  Technical progress is significant with the FDI path because the declining trend of technical levels is reversed in Eastern regions after 1994. Unfortunately, this is not the case in both Middle and Western regions. The total factor productivity growth with the FDI path is attributed mainly to technical progress rather than efficiency improvements in both Eastern and Middle regions. However, neither technical progress nor efficiency improvement is found in Western regions (Chap. 5).
  Strategies are presented through transplantations of Eastern technical efficiency with the FDI path into Middle and Western regions, which not only succeed in raising technical efficiency, but also keep technical efficiency from decreasing (Chap. 6).
- *Sustainable efficiency with urbanization*
  Transplantation strategies with an interior migration of population by 30 % and a selected enhancement of investment by 30 % will be two promising strategies to reach balanced urbanization development between middle and eastern regions. Transplantation strategies with an additional 10 % rise in investment put into big cities with urbanization level than the overall average urbanization level and only changes 50 % of the original added value of investment in big cities to promote the overall urbanization level for sustainable efficiency (Chap. 7).

Chapter 1
Introduction

This book focuses on the identification and measuring of changes in production efficiency in China for sustainable growth. The introduction in this chapter is mainly presented to illustrate motivation, objectives and outline of the book.

1.1 Motivation

Economic growth is a perpetual topic, and only those who keep rapid and sustainable growth can survive fierce competitions in the world. As far as a miraculous economic growth is concerned, it shifts one country to another as a domino effect in the world.

- Japan embodied a miraculous growth in the 1960s;
- Brazil witnessed its highest growth in the 1970s;
- Korea, Singapore, Hong Kong and Taiwan, Asia's "four little dragons", experienced their rapid growth which drew worldwide attention in the 1980s;
- China's economy has been developing with surprising rapidity since the 1990s, averaging almost 10 % per year.

Table 1.1 presents the top 10 economies that witness highest economic growth speed in the world during 1978–2005, illustrating their GDP annual growth rates.

Figure 1.1 presents the GDP growth in china during 1978–2006, after two decades of two-digit economic growth, China has climbed to the fourth place in terms of its gross economy scale. The issue of sustainable development in China has been debated with worldwide interests. In 1997, Nobel Prize Laureate, Laurence R. Klein expressed his doubt on whether the fabulous growth trend in the past 19 years would continue for the following 20 or 30 years. Fogel (2006) has stated that China is expected to have done pretty well in achieving its growth despite of its various problems that stand in the way of China's economic improvement. Klein and Mak (2007) express their confidence in the sustainability of China's economic expansion. General Secretary of China Reform Foundation Gang Fan addressed at the 2007 Forum of Supply Chain Management in China that the fast growth in China

B. Xu et al., *Changes in Production Efficiency in China*,
DOI: 10.1007/978-1-4614-7720-4_1,

Table 1.1 Top 10 economies with highest growth during 1978–2005 (%)

Country	China	Botswana	Singapore	Taiwan, China	Korea, Rep.
Annual growth	9.73	8.04	7.08	6.79	6.66
Country	Vietnam (1985–2005)	Malaysia	Thailand	Bhutan (1981–2005)	Belize
Annual growth	6.66	6.37	6.12	6.08	5.97

Data source World Development Indicators (various years)

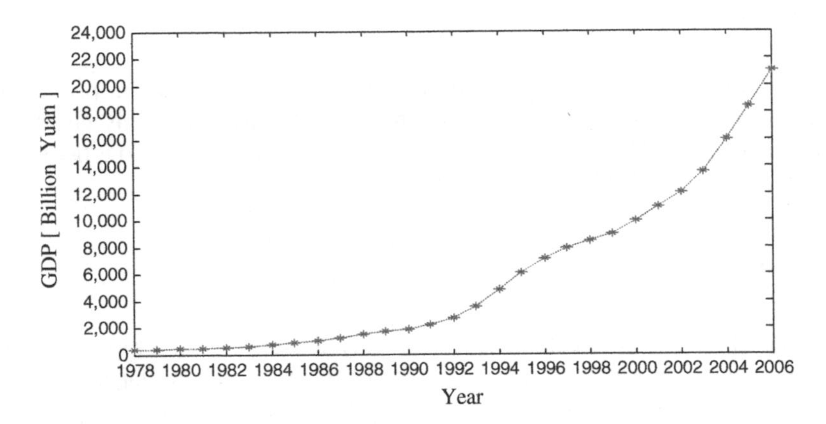

Fig. 1.1 Chinese GDP growth during 1978–2006

will continue without any sudden slowdown. The sustainable growth can be achieved with reliance on the reform effect, opening-up, education and technology, and urbanization process in China.

The history of China's growth reveals that the majority of growth is mostly attributed to large scale of investment. Specifically, the proportion of fixed asset investment to GDP has witnessed a fast rise of from 42 to 47 and even to 50 % during 2008–2009 and 2010–2011 in China. The fact naturally raised questions involving sustainable growth in China. In this sense, the key point to settle the debate lies in whether the behavior of investment, scale of investment, scale effect, allocation effect and technical transfer effect will be sustainable in production system in future. It is well accepted that both the investment behavior and investment scale will not last forever in any growth system.

Therefore, the questions remains to be explored are whether its resulting effects (scale effect, allocation effect and technical transfer effect) of investment will last for a long period. Specifically, first, whether the investment brought changes in production efficiency in the past China? Second, whether the promoting role of investment, especially foreign direct investment (FDI), on economic growth in Eastern China can be transferred into the growth in both Middle and Western China in future?

1.1.1 Present Framework

The production function provides the neoclassical model to investigate the growth,

$$Y_0 = F_0(A, K, L) \tag{1.1}$$

where Y_0 is the output, K and L are capital and labor, respectively, and A is the technical level.

The major interest of changes in production efficiency lies in identifying and measuring the efficiency change with a new factor P introducing into initial neoclassical production model (1.1). Specifically, the extended production function with the new factor is given by:

$$Y = F(A, K, L, P) \tag{1.2}$$

To identify and measure the changes of production efficiency due to new factor P, it is necessary to settle two questions.

The first one is how to identify the changes in production efficiency due to new factor P. The second one is how to make decision on optimal allocation in order to realize production efficiency of new factor P.

Take the total differential of model (1.2) with respect to P and it results in the following equation:

$$\frac{dY}{dP} = \frac{\partial F}{\partial P} + \frac{\partial F}{\partial K}\frac{dK}{dP} + \frac{\partial F}{\partial L}\frac{\partial L}{\partial P} + \frac{\partial F}{\partial A}\frac{dA}{dP} \tag{1.3}$$

Model (1.3) shows the output change due to new factor P can be split into three parts:

(a) Direct effect of factor P by $\partial F/\partial P$;
(b) Indirect effect of factor P via endogens variables K and L by $\partial F/\partial K$ and $\partial F/\partial L$;
(c) Indirect effect of factor P via exogenous technical variable by $\partial F/\partial A$.

The following analysis presents detail decomposition of the three parts.

1. Direct effect of factor P

It is unavailable to calculate $\partial F/\partial P$ directly since the underlying production function F is unknown. Take the first-order difference of Eq. (1.3) and it comes to the following linear model

$$\Delta Y = \frac{\partial F}{\partial P}\Delta P + \frac{\partial F}{\partial K}\Delta K + \frac{\partial F}{\partial L}\Delta L + \frac{\partial F}{\partial A}\Delta A + error \tag{1.4}$$

The technical level A is assumed to be exogenous and unknown. To replace technical level A, the existing research often uses a control factor C based on the endogenous growth theory, given by

$$\Delta Y = \frac{\partial F}{\partial P}\Delta P + \frac{\partial F}{\partial K}\Delta K + \frac{\partial F}{\partial L}\Delta L + \frac{\partial F}{\partial C}\Delta C + error \quad (1.5)$$

The empirical research on allocation efficiency of new factor P, i.e. $\partial \hat{F}/\partial P$, is often obtained by ordinary least square (OLS) regression after the unit root test and Granger causality test.

If $\partial \hat{F}/\partial P \geq 0$, the new factor P brings about positive production efficiency. Otherwise, if $\partial \hat{F}/\partial P < 0$, it is taken as the negative efficiency of factor P.

2. Indirect effect of factor P via endogenous capital and labor

The new factor P will indirectly impact production efficiencies of capital and labor by $\partial F/\partial K$ and $\partial F/\partial L$ via dK/dP and dL/dP in (1.3) respectively. Take the total differential of model (1.2) with respect to time t and get

$$\frac{dY}{dt} = \frac{\partial F}{\partial K}\frac{dK}{dt} + \frac{\partial F}{\partial L}\frac{dL}{dt} + \frac{\partial F}{\partial P}\frac{dP}{dt} + \frac{\partial F}{\partial A}\frac{dA}{dt} \quad (1.6)$$

Then

$$\begin{aligned}\frac{dY/dt}{Y} &= \frac{\partial F}{\partial K}\frac{K}{F}\frac{dK/dt}{K} + \frac{\partial F}{\partial L}\frac{L}{F}\frac{dL/dt}{L} + \frac{\partial F}{\partial P}\frac{P}{F}\frac{dP/dt}{P} + \frac{\partial F}{\partial A}\frac{A}{F}\frac{dA/dt}{A} \\ &\equiv \alpha\frac{dK/dt}{K} + \beta\frac{dL/dt}{L} + \lambda\frac{dP/dt}{P} + \frac{\partial F}{\partial A}\frac{A}{F}\frac{dA/dt}{A}\end{aligned} \quad (1.7)$$

where $\alpha = \frac{\partial F}{\partial K}\frac{K}{F} = \frac{\partial \ln F}{\partial \ln K}$ and $\beta = \frac{\partial F}{\partial K}\frac{L}{F} = \frac{\partial \ln F}{\partial \ln L}$ are the elasticity of capital and labor, respectively.

$\alpha + \beta$ is assumed to be the production scale. The estimation algorithm usually employs the logarithmic variable for analysis in order to obtain the stability of error item in model (1.5). Therefore,the $\partial F/\partial K$ and $\partial F/\partial L$ explain scale efficiency when taking the logarithm of original data.

The empirical investigation of scale efficiency is obtained by the OLS estimation $\hat{\alpha}$ and $\hat{\beta}$ by regression model (1.7) after replacing A by C, which indicate the values of α and β respectively.

3. Indirect effect of factor P via exogenous technical variable

The new factor P will indirectly impact production efficiency of technical variable A by $\partial F/\partial A$. The partial derivative $\partial F/\partial A$ is often obtained indirectly such as by "Solow Residual" or by decomposition of Malmquist Index (Kim and Park 2006) because A is exogenous and unavailable in observation data, given by

$$\frac{\partial F}{\partial A}\frac{A}{F}\frac{dA/dt}{A} \equiv \frac{dY/dt}{Y} - \alpha\frac{dK/dt}{K} + \beta\frac{dL/dt}{L} - \lambda\frac{dP/dt}{P} \quad (1.8)$$

Equation (1.8) presents the traditional empirical estimation of technical efficiency. More specifically, $(\partial F/\partial A)/(A/F)$ indicates the efficiency improvement and $(dA/dt)/A$ refers to the technical progress.

1.1.2 Questions

Given the research framework above, the empirical investigation provides us the following three questions due to uncontrolled factor C in (1.5), which is the second difficulty mentioned above. More specifically, the previous model framework brings about heated debate on whether factor input P changes production efficiency including allocation efficiency, scale efficiency and technical efficiency, which greatly impacts the decision-making of factor input strategy.

- **Changes in Allocation Efficiency of Resource**

The $\partial \hat{F}/\partial P$ in regression model (1.5), which is merely an average value, cannot express the dynamic efficiency of the changing P and would be either positive or negative. Another, because of different control factor C in (1.5), the analysis often brings about lots of contrary conclusions to the theoretical ones. For example, both positive and negative effects of public expenditure on economic growth are reported, and consequently raise the question on the allocation efficiency of public expenditure.

Question 1: Positive or negative effect of investment with the perspective of allocation efficiency?

The key to the question lies in the exploration of efficiency-oriented or equity-oriented allocation of investment strategy. Team (2004) points out the development of China's economy should emphasize more on the supply effect of fiscal policy. It is desirable to maintain high public capital expenditure in the economic primary take off period. Sun (2006) empirically illustrates that government spending shocks have *positive effects* on output, while tax revenues have negative effects. Similarly, Cao (2006) finds that the government expenditure scale has *positive* correlation with economic growth rate.

Conversely, Hansson and Henrekson (1994) hold that government transfers, consumption and total outlays have consistently *negative effects*, while educational expenditure has a positive effect, and government investment has *no effect* on private productivity growth. The impact is also found to work solely through total factor productivity and not via the marginal productivity of labor and capital. Chen and Yu (2006) point out that the raising proportion of public expenditure in GDP can *lower* the technological efficiency, but raising some parts, which to turn the public expenditure structure, of public expenditure in GDP can promote the technological efficiency. Ang (2008) finds that various public investment programs seem to have impacted *negatively* on economic development in Malaysia.

The debate of positive or negative effect on GDP results in different investment strategies under the framework of efficiency-oriented or equity-oriented allocation respectively.

- **Changes in Scale Efficiency of Factors**

The scale efficiency changes with *different control factor C* in regression model (1.5), which often puzzles the decision-making of investment strategy from the perspective of scale efficiency. For instance, the existing researches of FDI in China empirically confuse decision-makers with its investment strategy with contrary conclusions.

Question 2: Crowd-in or crowd-out effect of investment from the perspective of scale efficiency?

Wei (2002) holds that about 90 % of China's regional discrepancy in growth can be explained by the foreign direct investment (FDI) in Chinese authoritative *Economic Research Journal.* In the same Journal, Wu (2002) points out that less than 20 % of regional discrepancy is attributed to the regional differences of FDI.

Accordingly, Wei (2002) puts forward that energetic efforts should be made to actively absorb foreign capital into Middle and Western regions so as to push their developments; while Wu (2002) denies the possibility that the FDI regional distribution change could narrow the regional discrepancy in China. The indirect production efficiency of FDI depends on crowd-in and crowd-out effect on capital.

- **Changes in Technical Efficiency**

If the low production efficiency is attributed to little improvement in technical progress, the decision-making aiming to technical innovation is expected to be designed. If high rates of technical progress coexist with deteriorating technical efficiency, resulting in slow production efficiency, the decision-making aiming to bring improvements in learning-by-doing processes and in managerial practices (Kim and Han 2001) is expected to be adopted.

Question 3: Technical progress or efficiency improvement of technical variable from the perspective of technical efficiency?

Murakami (2007) demonstrates the entry of foreign-owned firms has a *positive* effect on the productivity of local firms in Japan as a result of *technology spillovers* in the long run. Fare et al. (2001) find that productivity growth is generally achieved through *technical progress*, and the efficiency change *negatively* contributes to productivity growth for Taiwanese manufacturing.

However, Cook and Uchida (2002) find that *efficiency improvement* dominates technical progress in developing countries. Lam and Shiu (2008) point the differences in efficiency scores are mainly due to the differences in the operating environments of different provinces, rather than the efficiency performance of telecommunications enterprises.

Kim and Park (2006) show both domestic and foreign R&D played an important role in increasing efficiency and technical progress in Korean manufacturing.

However, domestic R&D has more effect on *technical progress*, while foreign R&D has played a relatively stronger role in *efficiency improvement.*

Taken the perspective of technical transfer, whether the technical progress or efficiency improvement on economic growth in Eastern China can be transferred into the growth in both Middle and Western China is of great significance in order to narrow the regional discrepancy in China.

1.2 Objective

The objective of this book is to provide unique identification technique with empirical data during 1985–2006 to identify and measure the changes in production efficiency due to investment in cities or regions in China from the perspectives of allocation orientation, scale crowd-in effect and technical transfer. The innovative identification is established with application to implement optimal strategy of investment and migration for decision-making in urban planning through Path-Converged Design by means of production function.

First, it presents an observable benchmark as the criterion of the production efficiency to replace the production frontier surface with efficiency-oriented framework rather than output-oriented or input-oriented one. Second, the path-converged design is designed to select a controllable variable as a path of identification and to avoid uncontrollable natural variables.

The book launches specific researches from three parts:

1. Allocation efficiency orientation with investment and migration
   - Provides identification on allocation efficiency of investment and migration in cities, and on positive or negative effect on GDP growth in regions.
   - Implements simulation of optimal strategy on allocation efficiency of investment, and on migration allocation from inefficient cities to efficient ones and from regions with negative effect to regions with positive effect.
2. Crowd-in scale efficiency with foreign direct investment
   - Provides identification of crowd-out or crowd-in effect on scale efficiency in region using path identification of time-varying elasticity of production factors.
   - Implements simulation of alternative strategy on scale efficiency from regions with crowd-out effect to regions with crowd-in effect.
3. Technical efficiency transfer with investment
   - Provides evidence on whether technical efficiency realizes technical progress or efficiency improvement in regions.
   - Implements simulation of technical efficiency to change production efficiency in regions.

1.2.1 Innovative Approach

The objectives will be achieved by solving the two difficulties above.

- Presenting a **benchmark model** to replace production frontier surface. The idea of benchmark is derived from equilibrium of production structure in the perspective of management Science.
- Providing a **path model** to identify underlying production function. The idea of path is derived from sample path of stochastic process in the perspective of statistics.

The detailed information of the nonparametric identification technique is derived as follows. Usually, the research on identification has been broadly associated with the design of experiments. In biological and physical sciences, an investigator who wishes to make inferences about certain parameters can usually conduct controlled experiments to isolate relations. However, the cases in management engineering on social sciences are less fortunate. Certain facts can be observed and it is naturally required to arrange them in a meaningful way for studies. Yet their natural conditions usually can't be reproduced in laboratory. It is hard to control the variables and isolate the relations and it is almost impossible to identify an unknown latent system that produces data used in the analysis. However, fortunately in the view of management engineering on social sciences, the identification focuses on the change effects of the underlying structure generated by controlled conditions rather than the underlying structure itself. The steps here aim to build an identification framework of change effects due to the underlying structure.

Let x, y be a set of such observations. Structure S denotes the complete specification of probability distribution function, $F(x,y|S)$ of x and y. The set of all ***structure*** S is called a ***system*** $\mathbb{S}$. A possible prior structures S_0 of the ***system*** $\mathbb{S}$. is called a ***model***. The identification problem is to decide the judgment about ***model***, given structure S and observations x and y.

In the case of parametric identification, it is assumed that x and y are generated by parametric probability distribution function, that is

$$F(x,y|S_0) = F(x,y|\alpha) \tag{1.9}$$

where α is an m-dimensional real vector. Probability distribution function F is assumed known, which is conditional on α. However, parameter α is unknown. Hence, a structure is described by a parametric point α, and a model is described by a set of points $A \subset R^m$. Thus, the problem of distinguishing between structures is reduced to the problem of distinguishing between parameter points. In essence, this parameter identification can be solved through the estimation of unknown parameter α using statistical method.

In the case of nonparametric estimation, probability distribution function F is unknown. Given actual data x and y, the nonparametric estimations can be obtained for example, conditional density $f(y|x)$, regression function $m(x) = E(Y|x)$, etc.

The rationality of estimations is proved by the convergence of the nonparametric estimated $\hat{F}$ to the unknown F. In the case of nonparametric test, probability distribution function F is assumed to equal to a known function F_0, the nonparametric test is implemented by testing the null hypotheses $F = F_0$ versus $F \neq F_0$.

There has been nonparametric identification that only handles some dichotomous classification, such as the identification of regression models with misclassification on dichotomous regressor (Mahajan 2006; Chen et al. 2007), the nonparametric identification of the classical errors-in-variables model (Schennach et al. 2007) and semiparametric identification of structural dynamic optimal stopping time models (Chen 2007). Based on the ideas of parameter identification and nonparametric estimation, the book brings forward original nonparametric ***Path*** identification. Different from dichotomous classification, the identification strategy in this book focuses on the nonparametric path identification of an underlying structure.

Assume real-world data x, y are generated by nonparametric probability distribution function $F(x, y|S_0)$, where the probability distribution function F is unknown, and S_0 is called as an ***underlying structure.*** No parameters in F need to be estimated or dichotomous classification needs to be identified. In this case, it is impossible to reduce the problem of distinguishing between structures to the problem of distinguishing between parameter points. The concern is how to identify underlying structure S_0. The original identification of underlying structure S_0 with probability one can be achieved by the following three steps.

Step 1. Benchmark Model—to estimate underlying structure

Using the nonparametric approach, the unknown probability distribution function $F(x, y|S_0)$ can be estimated based on observed data x and y. For example, conditional density $f(y|x)$, regression function $m(x) = E(Y|x)$, etc.

Let $F_n(x, y|S_1)$ is an estimation of $F(x, y|S_0)$, where n is the sample size of (x, y), S_1 is called as a ***benchmark model.***

Intuitively, taken the view of statistics, the following assumption needs to be satisfied. The nonparametric estimation, $F_n(x, y|S_1)$, is the strong consistent estimation of $F(x, y|S_0)$, i.e.

$$P\Big(\lim_{n\to\infty} F_n(x, y|S_1) = F(x, y|S_0)\Big) = 1 \tag{1.10}$$

where P is a probability measurement.

Step 2. Path Model—to find controllable factors

Imitating the identification of parameter α in parameter identification, this study introduces an new observation variable Z into the probability distribution function $F(x, y)$ to obtain $F_n(x, y|Z)$. Recalling the law of diminishing marginal returns to Z *in view of economics*, and the law of large numbers of Z *in view of statistics,* it gives that restriction of a formation Z is indispensable. The variable Z is taken as a path model, if the following equation is satisfied:

$$P\Big(\lim_{n\to\infty}[F_n(x,y|Z) - F_n(x,y|S_1)] = 0\Big) = 1 \tag{1.11}$$

This equation in the book can be call as Z ***Path-Converged Design***. It is of crucial importance to the original nonparametric identification.

Compared with parameter identification of (4.3), it has ***Path model*** S_2 identifiable by path Z, i.e., there is $F_n(x,y|S_2)$, such that

$$F_n(x,y|S_2) = F_n(x,y|Z) \tag{1.12}$$

In other words, path structure S_2 can be found uniquely through observation variable Z.

According to parameter identification, the parameter α is unique. However, observation Z is unique only with probability one in (1.12). Hence, Z is not a unique identification of ***underlying structure*** S_0.

Compared with constant parameter α, observation Z may be a comprehensive integration set of various factors, which reflects the dynamic change of ***underlying structure*** S_0. Although the known observation Z is not a full identification of underlying structure S_0 itself, the Z identification focuses on the change effects generated by the underlying structure rather than the underlying structure itself. However, the change effects of S_0 identified by Z is not given explicitly since S_0 is unknown.

Step 3. Path Identification

Now, the problem of identification of underlying structure S_0 is reduced to the problem of identification of the change effects generated by the ***Path model*** S_2 and ***Benchmark model*** S_1.

Through the path Z, the identification of underlying structure $F(x,y|S_0)$ is reduced to the identification of change effects between benchmark model $F_n(x,y|S_1)$ and path model $F_n(x,y|S_2)$.

In fact, combine Eqs. (1.12) with (1.10) and get

$$P\Big(\lim_{n\to\infty} F_n(x,y|Z) = F(x,y|S_0)\Big) = 1$$

Hence, according to parameter identification, underlying structure S_0 is identifiable by the observed path Z, with probability one.

The following definition states the nonparametric Z path identification with probability one.

Definition 1 Underlying structures S_0 is said to be Z ***path identifiable*** with probability one, if the following equations are satisfied:

$$\text{(a)} \quad P\Big(\lim_{n\to\infty} F_n(x,y|S_1) = F(x,y|S_0)\Big) = 1$$

$$\text{(b)} \quad P\Big(\lim_{n\to\infty}[F_n(x,y|S_2) - F_n(x,y|S_1)] = 0\Big) = 1$$

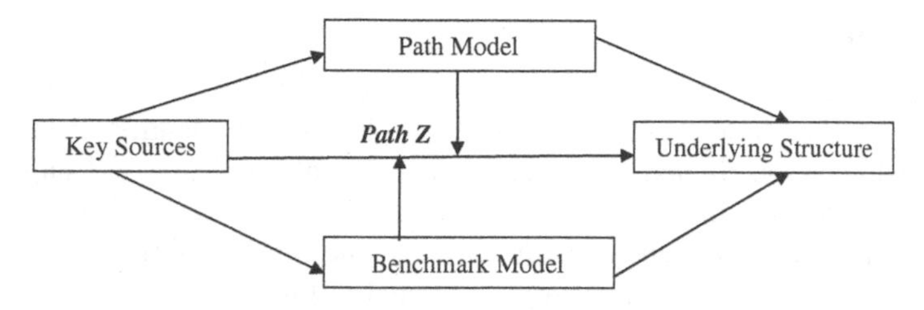

Fig. 1.2 Nonparametric ***Path*** identification

where $F_n(x, y|S_2) = F_n(x, y|Z)$. The identification means that path Z realizes the identification of underlying structure S_0 by observing the change effects of path model S_2 based on benchmark model S_1 with probability one.

Figure 1.2 illustrates the main idea of nonparametric path identification.

Two remarks are given considering the path identification framework in this book.

Remark Fine (2000) mentions the econometrics of endogenous growth has confronted with three problems, that is, the problems of time-varying feature, model selection and independence assumption. The time-varying elasticity production function introduced in this chapter (Xu and Watada 2007b) has settles the first problem. Given a benchmark structure of driven growth, the nonparametric identification of underlying structure is obtained by comparison of benchmark with path structures. Therefore, the identification handles the model selection issue by comparison with a given benchmark structure. Moreover, the comparison successfully avoids the issue of independence assumption.

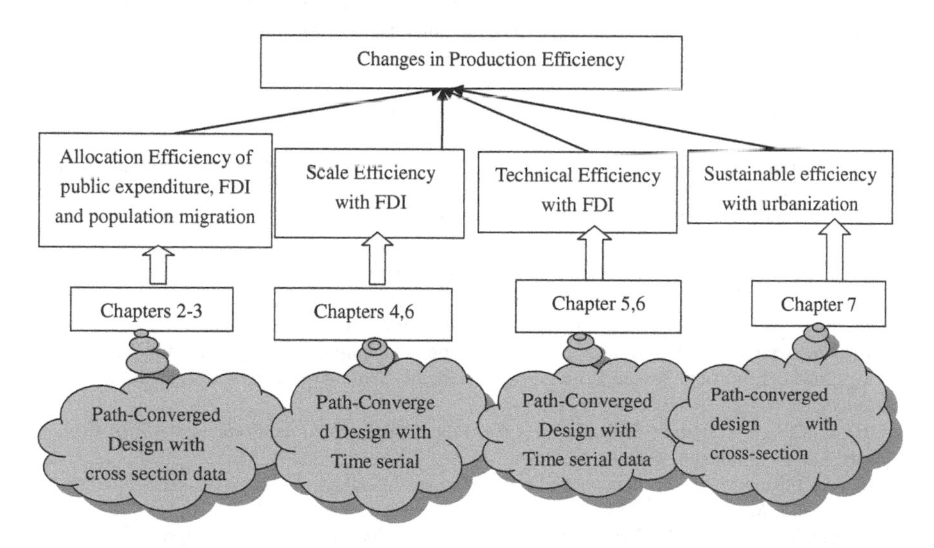

Fig. 1.3 The structure of the book

1.3 Outline

Path-converged design is established with application to identify and measure changes in allocation efficiency in Chaps. 2 and 3, scale efficiency in Chaps. 4 and 6, technical efficiency in Chaps. 5 and 6, sustainable efficiency with urbanization in Chap. 7, and with the general *path-converged design* approach in Chap. 4

The structure of the book is clearly presented in Fig. 1.3.

References

Ang, J. (2008). What are the mechanisms linking financial development and economic growth in Malaysia? *Economic Modelling, 25*(1), 38–53.

Cao, Y. (2006). An empirical study of the influences of public expenditure scale on GDP in China. *Finance and Trade Research, 17*(4), 65–68.

Chen, L. (2007), Semiparametric identification of structural dynamic optimal stopping time models, *Cemmap Working Paper*, CWP06/07.

Chen, X., & Yu, J. (2006). The influence analysis on public expenditure to the technique efficiency of China. *Frontiers of Economics in China, 1*(2), 296–310.

Chen, X., Hu, Y. & Arthur, L. (2007). Nonparametric identification of regression models containing a misclassified dichotomous regressor without instruments, *Cemmap working paper.*

Cook, P., & Uchida, Y. (2002). Productivity growth in East Asia: A reappraisal. *Applied Economics, 34*, 1195–1207.

Fare, R., Grosskopf, S., & Lee, W. F. (2001). Productivity and technical change: The case of Taiwan. *Applied Economics, 33*(15), 1911–1924.

Fine, B. (2000). Endogenous growth theory: A critical assessment. *Cambridge Journal of Economics, 24*, 245–264.

Hansson, P., & Henrekson, M. (1994). A new framework for testing the effect of government spending on growth and productivity. *Frontiers of Economics in China, 81*(3-4), 381–401.

Kim, S., & Han, G. (2001). A decomposition of total factor productivity growth in Korean manufacturing industries: A stochastic frontier approach. *Journal of Productivity Analysis, 16*, 269–281.

Kim, T., & Park, C. (2006). Productivity growth in Korea: Efficiency improvement or technical progress? *Applied Economics, 38*(8), 943–954.

Lam, P., & Shiu, A. (2008). Productivity analysis of the telecommunications sector in China. *Telecommunications Policy, 32*(8), 559–571.

Mahajan, A. (2006). Identification and estimation of regression models with misclassification. *Econometrica, 74*(3), 631–664.

Murakami, Y. (2007). Technology spillover from foreign-owned firms in Japanese manufacturing industry. *Journal of Asian Economics, 18*, 284–293.

Schennach, S. M., Hu, Y. & Lewbel, A. (2007). Nonparametric identification of the classical errors-in-variables model without side information, *Cemmap Working Paper,* CWP14/07.

Sun, L. (2006). Empirical research on dynamic effects of fiscal policy in China: 1998–2004. *Finance and Trade Research, 17*(1), 59–64.

Wei, H. K. (2002). Effects of foreign direct investment on regional economic growth in China. *Journal of Economic Research, 4*, 19–26.

Wu, J. (2002). Regional discrepancy of FDI in China and its effect on economic growth. *Journal of Economic Research 4*, 27–34.

Xu, B. and Watada, J. (2007b), A time-varying elasticity production function model with urbanization endogenous growth: Evidence of China, *©2007 IEEE, 0-7695-2882-1/07.*

Chapter 2
Changes in Allocation Efficiency from Equity to Efficiency-Oriented

How investment changes production efficiency puzzles world's economic growth all the time. The investment philosophy in most countries is oriented with priority in efficiency promotion. It does not deny the fact that investment itself is unsustainable to economic growth, only when efficiency promotion is realized at a certain level, the pursuit of equality will be considered for sustainable growth. While under the background of Eastern culture, the initial investment philosophy in China is majorly oriented with priority of equity allocation rather than efficiency promotion for growth. Specifically, both domestic governmental financial investment and private investment emphasize the equity oriented growth.

Since the opening-up in 1978, absorbing foreign direct investment (FDI) has seemed to be an effective way to realize growth. FDI, with its specific character of emphasizing efficiency promotion, has inevitably impacted on the initial investment philosophy of equity. With the combination of Chinese domestic investment and foreign direct investment, the investment philosophy in China has been forced to search a balance between efficiency promotion and equity pursuit. To search the balance, the transfer from equity pursuit to efficiency promotion, and eventually to a final balance, the production system has to experience changes in production efficiency from different aspects.

Being the pioneer during opening-up, large amount of private investment in Eastern region keeps an investment philosophy of efficiency promotion with the affects of FDI. Both the FDI absorption and domestic private investment show their very first direct effects on production by changes in resource allocation and hence to allocation efficiency in production system. This chapter will measure the changes in allocation efficiency with investment strategy based on macro-economic data with the index of governmental financial expenditure because the micro data from domestic private investment are unavailable.

B. Xu et al., *Changes in Production Efficiency in China*,
DOI: 10.1007/978-1-4614-7720-4_2,

2.1 Equity or Efficiency-Oriented Growth

To explore whether the growth is equity pursuit or efficiency promotion oriented, the chapter will start from the debate on how governmental public expenditure (PE) and FDI exerts impact on allocation production efficiency via the perspective of GDP distribution. Theoretically, the debate mostly concerned about whether the marginal direct effect of new factor P $\partial\hat{F}/\partial P$ is positive or negative in Eq. (1.4).

On the one hand, both positive and negative effects of public expenditure on economic growth are reported according to several existing empirical researchers. Consequently, the decision-making debate on different allocation strategy of public expenditure is raised under the allocation efficiency oriented framework.

On the other hand, the effect of FDI on economic growth still remains contentious. Blomstrom et al. (1994) have found that whether FDI can help contribute to the economic growth depends on the host country's developing level. Kawai (1994), Balasubr et al. (1996) argue that the opening level of the host country was vital to FDI's effects on economic growth. Borensztein et al. (1998) argue that FDI's effects on economic growth depend on the host country's human capital stock. Alfaro et al. (2004) have found that FDI's effects on economic growth hinge on the host country's financial development level. Goss et al. (2007) point out that foreign capital was responsible for more than one-third of the manufacturing sector's productivity growth in U.S. between 1988 and 1994. More importantly, foreign capital contributed almost 16 % of productivity growth over 1995 and 1999, significantly more than domestic capital's contribution.

Luo (2007) finds that direct effects of FDI on economic growth were of insignificance based on panel data at provincial level in China. However, through improving technical efficiency and "crowding in" domestic investment, FDI produced *positive effects* on China's economy. Surprisingly, Dierk et al. (2008) find that in the vast majority of countries, there exists neither a long-term nor a short-term effect of FDI on growth, which challenges the widespread belief that FDI generally has a *positive impact on* economic growth in developing countries.

Both positive and negative impacts of public expenditure and FDI on growth can be taken as the changes in allocation efficiency for growth since they exert as direct effect of investment on production growth in Chap. 1. Aiming to settle the debate, the study in this chapter attempts to explore how the domestic investment and FDI will change allocation efficiency in China from the perspective of macroeconomic data with following two questions.

The first one is how to identify the allocation efficiencies of public expenditure and FDI in cities, and the positive or negative effects of public expenditure and FDI on growth in regions.

The second one is how to realize allocation efficiencies of public expenditure and FDI with changes from inefficient cities to efficient ones and from negative impact regions to positive impact ones so as to make optimal decisions on investment.

To settle above two questions, this chapter establishes a Path-converged design to identify and realize the allocation efficiencies of FDI and PE in different regions of China. Different from previous studies, the new approach allows us to identify production efficiency from a perspective of dynamic distribution rather than the one of average impact parameters.

More specifically, the study firstly identifies the positive and negative impacts of PE and FDI on GDP per capita, secondly identifies the effective intervals with strengths of their allocations, and thirdly presents joint identification of allocation efficiency of PE and FDI, and finally realizes the allocation efficiencies of PE and FDI with changes from inefficient cities to efficient ones and changes from negative impacts to positive impacts due to PE and FDI in regions. The study in this chapter is a further research based on the study of Xu and Watada (2008c).

The rest of this chapter is organized as follows. Section 2.2 establishes the Path-converged design of allocation efficiency. Section 2.3 identifies the single and joint allocation efficiencies of FDI and PE. Section 2.4 measures the allocation efficiency promotion of the investment strategies of PE. Section 2.5 measures the allocation efficiency promotion of the investment strategies of FDI. Section 2.6 finally presents the conclusion.

2.2 Path-Converged Design of Allocation Efficiency

According to the research framework in Chap. 1, both domestic financial expenditure and FDI promote the growth through three channels (Luo 2007) based on the neoclassical growth theory (Solow 1957) and the endogenous growth theory (Romer1986; Lucas 1988). First, as investment, PE or FDI can have direct effect on economic growth through $\partial F/\partial P$. Secondly, by influencing domestic capital accumulation and labor participation, PE or FDI can have indirect effect on economic growth through the changes in $\partial F/\partial K$ and $\partial F/\partial L$. Thirdly, PE or FDI can exert indirect impact on economic growth through affecting technological efficiency in host countries through the changes in $\partial F/\partial A$. The research in this chapter focuses on identification of changes in production efficiency due to PE and FDI.

However, both domestic public expenditure and FDI are dynamic and impossible to be linear in changes of production efficiency. The traditional linear models are improper to depict the changes in allocation efficiency due to PE and FDI. Guided by dynamic identification, this chapter introduces a probability distribution to character the dynamic underlying structure which is constructed by the real-world GDP per capita observation. The identification refers to the inferences drawn from the probability distribution of the observed variables.

Let $X_i = X_i(t),\ i = 1, \cdots, n$, be the GDP per capita observation of city i at time t, Underlying structure of the overall GDP per capita is supposed as:

$$Y_o = F(A, K, L)$$

To investigate allocation efficiency,the underlying density function of underlying structure is given

$$S_o : f(x,\ X) \tag{2.1}$$

where $X = (X_1, X_2, \cdots, X_n)$.

The estimation of underlying density function of GDP per capita is given by kernel density estimation as follows.

$$S_1 : f_n(x,\ X) = \frac{1}{n}\sum_{i=1}^{n} h^{-1}K\left(\frac{x - X_i}{h}\right) \tag{2.2}$$

where $K(\cdot)$ stands for kernel function and h the bandwidth.

Both the investments of FE and FDI are guided by governmental policies; therefore, this chapter takes both FE and FDI as institutional variables for analysis, and furthermore takes as possible investment strategies to change in allocation efficiency.

To identify the initial allocation efficiency of PE and FDI in different regions of China, conditional kernel density approach of underlying structure of GDP per capita is provided with Path-converged design.

Conditional structure

$$Y_1 = F(A, K, L, FDI)$$

Estimation of conditional density function of conditional structure is given

$$S_2 : f_n(x,\ X|FDI) = \frac{f_n(x,\ X,\ FDI)}{f_n(FDI)} = \frac{\frac{1}{n}\sum_{l=1}^{n} h^{-2}K_0\left(\frac{x-X_l}{h},\ \frac{FDI-FDI_l}{h}\right)}{\frac{1}{n}\sum_{j=1}^{n} h^{-1}K_1\left(\frac{FDI-FDI_j}{h}\right)} \tag{2.3}$$

Take a production kernel

$$K_0\left(\frac{x - X_1}{h},\ \frac{FDI - FDI_l}{h}\right) = K_1\left(\frac{FDI - FDI_l}{h}\right)K\left(\frac{x - X_1}{h}\right) \tag{2.4}$$

Let

$$\omega_l = \frac{K_1\left(\frac{FDI-FDI_l}{h}\right)}{\sum_{j=1}^{n} K_1\left(\frac{FDI-FDI_j}{h}\right)} \tag{2.5}$$

Here $\omega_i = \omega_i(t)$ is a weight of the FDI or FE of city i at time t, $i = 1, \cdots, n$. Therefore, conditional density function is given by

$$f_n(x, X, \omega) = f_n(x, X|FDI) = \sum_{j=1}^{n} \omega_j h^{-1}K\left(\frac{x - X_i}{h}\right) \tag{2.6}$$

Technically, the identification of allocation efficiency of FDI is difficult because the density $f_n(x, X, \omega)$ varies with different ω. The same situation occurs when it come to the identification of control factor C in (1.5). Guided by the identification in Chap. 1, the identification of allocation efficiency of FDI in (2.6) can be realized by the difference between Eqs. (2.1) and (2.6) as follows,

$$f_n(x, X, \omega) - f(x, X) \tag{2.7}$$

Decomposing

$$f_n(x, X, \omega) - f(x, X) = (f_n(x, X, \omega) - f_n(x, X)) + (f_n(x, X) - f(x, X)) \tag{2.8}$$

Taken the perspective of Statistics, with the selected ω, the optimal convergence rate $\lim_{n \to \infty} a_n = 0$ as below holds true with probability one.

$$f_n(x, X, \omega) - f(x, X) = a_n$$

However, FDI investment strategy is derived from production allocation efficiency rather than optimal convergence rate.

Recalling the law of diminishing marginal efficiency of $FDI_n = FDI(t, n)$ from the perspective of Economics, it gives

$$P\left\{\lim_{t \to \infty} \frac{\partial FDI(t, n)}{\partial t} = 0\right\} = 1 \tag{2.9}$$

Recalling the law of large numbers of $FDI(t,\ n)$ from the perspective of Statistics, it has

$$P\left\{\lim_{n \to \infty} \frac{1}{n} \sum_{j=1}^{n} (FDI(t,\ j) - EFDI(t,\ j)) = 0\right\} = 1 \tag{2.10}$$

Equations (2.9) and (2.10) indicate that FDI (or the average level of FDI) will be a constant when $t \to \infty$ (or $n \to \infty$).

The theories in both Economics and Statistics do not present specific function of $P(t,\ n)$. However, they bring about the analysis in the view of engineering as follows. The innovative *Path convergence design* approach in this chapter is built on the change in allocation efficiency through the following three steps.

Step 1. Benchmark Model

Rewrite (2.2) and get the equilibrium structure S_1 by:

$$f_n(x, X) - \frac{1}{n} \sum_{i=1}^{n} h^{-1} K\left(\frac{x - X_i}{h}\right) = 0 \tag{2.11}$$

Intuitively, *taken the perspective of statistics,* as estimation of underlying structure the following assumption needs to be satisfied:

$$P\left(\lim_{n\to\infty} f_n(x, X(t)) = f(x, X(t))\right) = 1 \tag{2.12}$$

And

$$P\left(\lim_{t\to\infty} f_n(x, X(t)) = f(x, X(t))\right) = 1 \tag{2.13}$$

Assumption (2.12) and (2.13) indicate that $f_n(x, X)$ is a strong consistent estimator of $f(x, X)$.

Step 2. Path Model

Suppose a new exterior factor FDI enters into the equilibrium structure S_1, the change process from equilibrium production structure S_1 to the disequilibrium production structure S_2 is presented as follows:

$$f_n(x, X, \omega) - \frac{1}{n}\sum_{i=1}^{n} h^{-1}K\left(\frac{x - X_i}{h}\right) \neq 0 \quad f \quad \omega_j \neq\equiv n^{-1}, \text{ for any } j \tag{2.14}$$

Since FDI is a new factor that introduced in the initial equilibrium structure, the change process from equilibrium to disequilibrium structure is due to the new factor FDI. To depict the change process, rewrite (2.6) and get path model S_2:

$$f_n(x, X, \omega) - \sum_{i=1}^{n} \omega_i h^{-1}K\left(\frac{x - X_i}{h}\right) = 0 \tag{2.15}$$

Step 3. *Path-converged design*

Models (2.11) to (2.14) describe the change process from equilibrium production structure to disequilibrium ones. (2.16) and (2.17) imply that the disequilibrium production structure (2.14) will adjust to re-equilibrium production structure.

Now, the research designs a $fdi_j = fdi_j(t)$, a function of $FDI(t, j)$, to replace $FDI(t, j)$, such that re-equilibrium as following

$$P\left(\lim_{n\to\infty} [f_n(x, X(t), \omega^*(t)) - f_n(x, X(t))] = 0\right) = 1 \tag{2.16}$$

And

$$P\left(\lim_{t\to\infty} [f_n(x, X(t), \omega^*(t)) - f_n(x, X(t))] = 0\right) = 1 \tag{2.17}$$

where $\omega^*(t)$ definition see (2.19) below and it replace $\omega(t)$ in (2.5).

Replace FDI_t by $fdi_j(t) = FDI_j(t)/GDP_j(t), \quad t = 2, \cdots,$

Now the definition of K_1 in (2.3) is presented as follows.

Taking

$$\lim_{x \to \infty} K_1(x) = c \tag{2.18}$$

where c is a constant.

Decomposing

$$\omega^*(l, n) = \frac{K_1\left(\frac{fdi - fdi_l}{h}\right)}{\sum_{j=1}^{n} K_1\left(\frac{fdi - fdi_j}{h}\right)} = \frac{K_1\left(\frac{fdi - fdi_l}{h}\right)\{I(fdi = fdi_j) + I(fdi \neq fdi_j)\}}{\sum_{j=1}^{n} K_1\left(\frac{fdi - fdi_j}{h}\right)\{I(fdi = fdi_j) + I(fdi \neq fdi_j)\}}$$

Now, if $fdi = fdi_i$, for some l, then

$$K_1\left(\frac{fdi - fdi_i}{h}\right) = K_1(0)$$

Thus

$$\omega^*(l, n) = \frac{K_1\left(\frac{fdi - fdi_l}{h}\right)}{\sum_{j=1}^{n} K_1\left(\frac{fdi - fdi_j}{h}\right)} = \frac{K_1\left(\frac{fdi - fdi_l}{h}\right) I(fdi \neq fdi_j)}{\sum_{j=1}^{n} K_1\left(\frac{fdi - fdi_j}{h}\right) I(fdi \neq fdi_j)}$$

if $fdi \neq fdi_i$,

$$\lim_{n \to \infty} \omega^*(l, n) \leq \lim_{n \to \infty} \frac{c + \varepsilon}{mc_0 + (n - m)(c - \varepsilon)} = 0 \tag{2.19}$$

where $m < n$, such that

$$c - \varepsilon \leq K_1\left(\frac{fdi - fdi_l}{h}\right) \leq c + \varepsilon$$

and

$$c_0 = \min_{1 \leq l \leq m} K_1\left(\frac{fdi - fdi_l}{h}\right)$$

for some $0 \leq \varepsilon \leq c$, h small enough, provided $\lim_{x \to \infty} K_1(x) = c$.

Equation (2.19) implies

$$P\left(\lim_{t \to \infty}\left[\omega^*(t, n) - \frac{1}{n}\right] = 0\right) = 1 \tag{2.20}$$

which implies (2.17) holds true according to the definition of (2.2) and (2.6).

The proof of (2.16) is similar to the proof of Theorem 1 below. Here $\omega^*(t, n)$ is called as *Path-converged design*.

Applications of (2.12), (2.13), (2.16), and (2.17) to (2.7), the identification of FDI investment efficiency is obtained by:

$$f_n(x, X, \omega^*) - f_n(x, X) \tag{2.21}$$

Now the *Path* identification of underlying density function is given as follows.

Definition 1 Underlying density function S_0 is said to be ω *path identifiable* with probability one, if the following equations are satisfied:

$$P\left(\lim_{t\to\infty} f_n(x, X(t)) = f(x, X(t))\right) = 1$$

$$P\left(\lim_{n\to\infty} f_n(x, X(t)) = f(x, X(t))\right) = 1 \tag{2.22}$$

and

$$P\left(\lim_{t\to\infty} [f_n(x, X(t), \omega(t)) - f_n(x, X(t))] = 0\right) = 1$$

$$P\left(\lim_{n\to\infty} [f_n(x, X(t), \omega(t)) - f_n(x, X(t))] = 0\right) = 1 \tag{2.23}$$

The path identification here realizes not only identification of the resource allocation efficiency along a path ω,but also identification of change of production efficiency of the underlying density function S_0 by observing the change efficiency of path model S_2 based on benchmark model S_1 with probability one.

Definition 2 The city x is said to be ω allocation inefficient if $f_n(x, X, \omega) < f_n(x, X)$ and S_0 is ω *path identifiable*. Furthermore, the ω inefficient interval is comprised of all inefficient points; i.e., inefficient interval $\bar{D} = \{x : x \text{ is ineficiency}\}$.

Definition 3 The strength of ω path inefficiency is defined by the integral area between the benchmark density function and the path density function at an inefficiency interval, i.e. $\int_D (f_n(x, X) - f_n(x, X, \omega))dx$ if the city x is ω allocation inefficient.

The inefficiency strength indicates the amount of investment for reallocation strategy.

For large samples, it is known that the nonparametric estimation is not sensitive to the different choices of kernel functions (Silverman 1986). The kernel function employed in this study is Gaussian kernel function:

$$K(u) = \frac{1}{\sqrt{2\pi}} \exp\left(-\frac{u^2}{2}\right) \tag{2.24}$$

Bandwidth h is selected by a Cross-Validation Approach (Stone 1984).

Random variables $X_1, X_2, \cdots, X_n$ are said to be strongly mixing (Kim and Lee 2005) if $\lim_{n\to\infty} \alpha(n) = 0$, where

$$\alpha(n) = \sup_{k\geq 1} \alpha(F_k, G_{k+n}) = \sup_{k\geq 1} \sup_{(A,B)\in F_k\times G_{k+n}} |P(A\cap B) - P(A)P(B)|$$

where $F_k = \sigma(X_i, i \leq k)$, $G_t = \sigma(X_i, i \geq t)$ and $\sigma(\cdot)$ denotes σ-field, it is generated by subset of $X_1, X_2, \cdots, X_n$.

Let X_i, $i = 1, \cdots, n$, be the GDP per capita observation of city i. Note the fact that if $|X_i - X_j| \leq h$, h is some constant, then maybe similarity of resource allocation in neighbor cities between i and j. Hence suppose X_i, $i = 1, \cdots, n$ strongly mixing is reasonable.

Theorem 1 *Suppose* $X_1, X_2, \cdots, X_n$ *be strongly mixing sequence,and* $\sum_{n=1}^{\infty} \alpha(n) < \infty$, *then underlying density function* $f(x, X)$ *is* ω *path identifiable.*

Proof In terms of (2.12), (2.13) and (2.17), it only needs to prove

$$P\Big(\lim_{n\to\infty} [f_{\omega n}(x) - f_n(x)] = 0\Big) = 1 \tag{2.25}$$

where $f(x) = f(x, X)$, $f_n(x) = f_n(x, X)$, $f_{\omega n}(x) = f_n(x, X, \omega)$.

Using the studies of Kim and Lee (2005), it is easy to get

$$P\Big(\lim_{n\to\infty} [f_n(x) - f(x)] = 0\Big) = 1 \text{ and } P\Big(\lim_{n\to\infty} [f_{\omega n}(x) - f(x)] = 0\Big) = 1$$

provided

$$\sum_{l=1}^{n} \omega_l = \sum_{l=1}^{n} \frac{K_1\left(\frac{fdi - fdi_l}{h}\right)}{\sum_{j=1}^{n} K_1\left(\frac{fdi - fdi_j}{h}\right)} = 1$$

It has

$$\begin{aligned}
&P\Big(\lim_{n\to\infty} [f_{\omega n}(x) - f_n(x)] \neq 0\Big) \\
&= P\Big(\lim_{n\to\infty} [f_{\omega n}(x) - f(x)] + \lim_{n\to\infty} [f(x) - f_n(x)] \neq 0\Big) \\
&= P\Big(\lim_{n\to\infty} [f_{\omega n}(x) - f(x)] + \lim_{n\to\infty} [f(x) - f_n(x)] \neq 0,\ \lim_{n\to\infty} [f(x) - f_n(x)] = 0\Big) \\
&\quad + P\Big(\lim_{n\to\infty} [f_{\omega n}(x) - f(x)] + \lim_{n\to\infty} [f(x) - f_n(x)] \neq 0,\ \lim_{n\to\infty} [f(x) - f_n(x)] \neq 0\Big) \\
&\leq P\Big(\lim_{n\to\infty} [f_{\omega n}(x) - f(x)] \neq 0,\ \lim_{n\to\infty} [f(x) - f_n(x)] = 0\Big) \\
&\quad + P\Big(\lim_{n\to\infty} [f(x) - f_n(x)] \neq 0\Big) \\
&\leq P\Big(\lim_{n\to\infty} [f_{\omega n}(x) - f(x)] \neq 0\Big) + P\Big(\lim_{n\to\infty} [f(x) - f_n(x)] \neq 0\Big) = 0
\end{aligned}$$

Hence (2.25) is true. Now, underlying density function S_0 is identifiable through path ω with probability one.

2.3 Identification of Allocation Efficiency

The following analyses present the identification of existing allocation efficiency of PE and FDI in different regions of China through GDP per capita distribution. The weighted kernel density estimation serves as main approach to realize the path identification, including the FDI, PE, and joint FDI and PE weighted kernel density estimation, which are called as FDI path, PE path, and joint path identification for short respectively.

2.3.1 Data Description

The collected data sets are the cross-city data of prefecture-level cities in China in 2006. The data sets contain GDP, foreign direct investment (actually used), and local public budgetary expenditure for each city. The data are available at http://newibe.cei.gov.cn/, China Urban Statistical Yearbook 2007. According to the classification criterion of China City Statistical Yearbook 2007, *Eastern region* contains Beijing, Tianjin, Hebei, Liaoning, Shanghai, Jiangsu, Zhejiang, Fujian, Shandong, Guangdong and Hainan province; *Middle region* contains Shanxi, Neimenggu, Jilin, Heilongjiang, Anhui, Jiangxi, Henan, Hubei, Hunan; and *Western region* contains Guangxi, Chongqing, Sichuan, Guizhou, Yunnan, Xizang, Shanxi, Gansu, Ningxia, Qinghai and Xinjiang. In the study, 254 prefecture-level cities are included, with 101 eastern cities, 104 middle cities and 49 western cities.

2.3.2 Single Identification of Allocation Efficiency of FDI and PE

Figure 2.1 presents the path identification of GDP per capita distribution of all prefecture-level cities in China in 2006. The horizontal axis is the standardized GDP per capita, which is the division of a city's actual GDP per capita by national GDP per capita in 2006. It illustrates the FDI path identification shifts rightward while the PE path identification leftward compared with benchmark identification of GDP per capita distribution. The initial public expenditures are allocation efficiency under the equity oriented investment philosophy. However, FDI is allocation efficiency while public expenditure is allocation inefficiency for overall China under efficiency oriented investment philosophy. The transfer from equity pursuit to efficiency promotion is the first step to final balance of investment

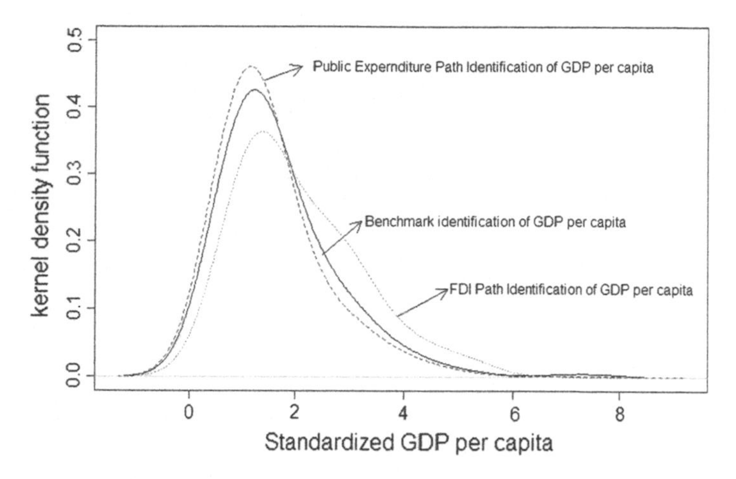

Fig. 2.1 Path identification of GDP per capita distribution in China in 2006

Table 2.1 Average levels (EX) of GDP per capita with path identifications for each region

Identification region	Benchmark	FDI path	PE path
Eastern region	2.218	2.515 (+)	2.060 (−)
Middle region	1.414	1.551 (+)	1.249 (−)
Western region	1.140	1.205 (+)	1.073 (−)
Overall region	1.685	2.041 (+)	1.522 (−)

strategy. Therefore, the following study will carry on the identification and measuring with efficiency-oriented framework.

The average levels of GDP per capita (expected values of GDP per capita distribution) are 1.685, 2.041 and 1.522 respectively for benchmark identification, FDI path identification and PE path identification, which validates FDI has positive impact and public expenditure has negative impact on production efficiency. And obviously, the positive impact of FDI on GDP per capita is stronger than the negative impact of PE on GDP per capita.

Note: the average levels of GDP per capita are presented by expected values of GDP per capita distributions. For example, $EX_1 = \int_0^{\infty} xf_n(x, X)dx$ is the average level of GDP per capita distribution with benchmark identification. The figures in Table 2.1 illustrate that FDI path increases and public expenditure decreases the average levels of GDP per capita, implying FDI has positive impact and public expenditure has negative impact on production efficiency in all regions.

Figures 2.2, 2.3, 2.4 are path identifications of GDP per capita distribution in Eastern, Middle and Western China in 2006. Similar to the path identification of GDP per capita distribution in China, the path identifications in Eastern, Middle and Western China witness positive impact of FDI and negative impact of PE on GDP per capita distribution, which is strongly demonstrated by the average levels of GDP per capita with the path identifications for each region in Table 2.1.

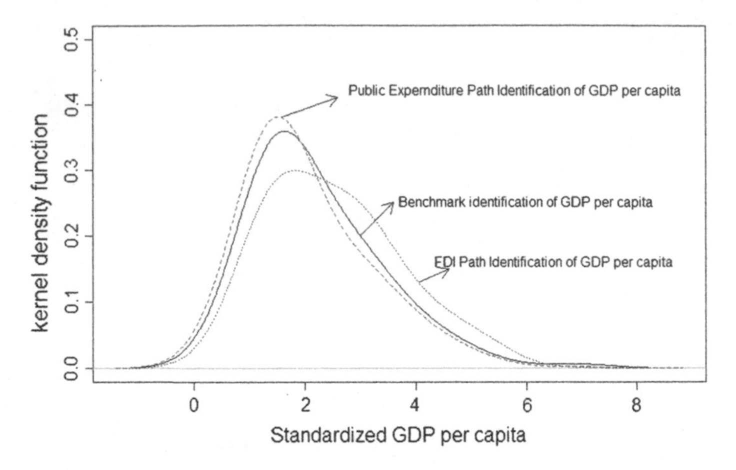

Fig. 2.2 Path identification of GDP per capita distribution in Eastern China in 2006

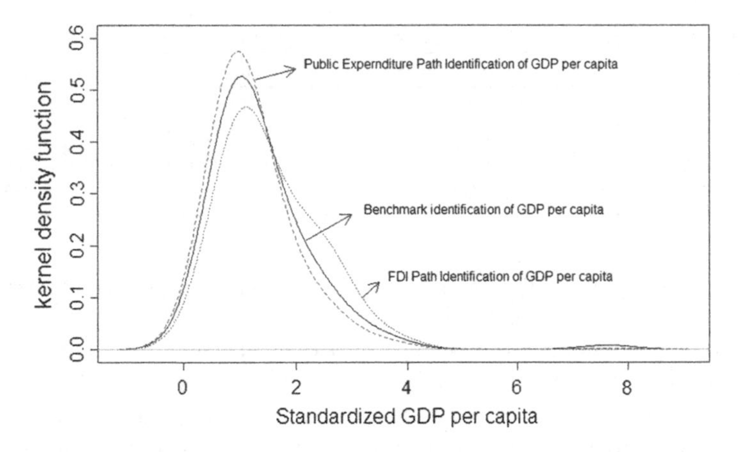

Fig. 2.3 Path identification of GDP per capita distribution in Middle China in 2006

2.3.3 *Effective Interval with Strengths of FDI and PE*

To explore the allocation efficiencies of FDI and PE in different Chinese regions in depth, the empirical results are two-fold.

On the one hand, the path identifications clarify *investment strategies* on where to invest FDI and public expenditure is more effective, the effective intervals are given in Table 2.2. In the study, the effective investment interval of FDI or PE defined by the intervals where the curve of FDI or PE path model is higher than one of benchmark model, where the investment in the area realizes allocation efficiency. First, the effective investment area of public expenditure locates at interval (0, 1.690) of standardized GDP per capita because the PE path

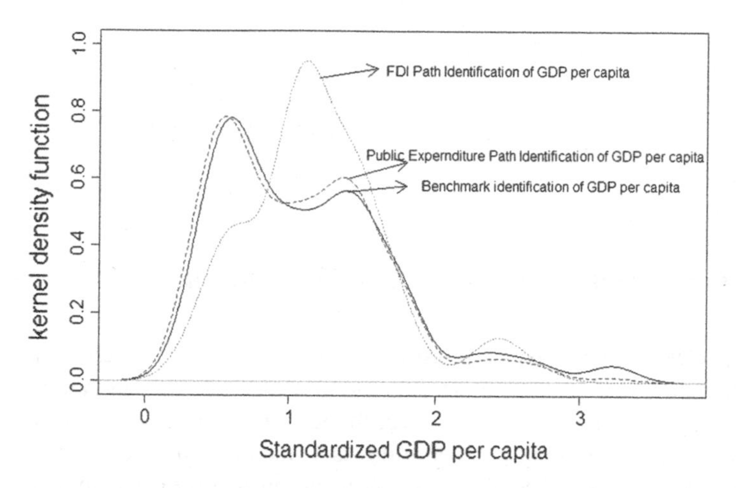

Fig. 2.4 Path identification of GDP per capita distribution in Western China in 2006

Table 2.2 Effective intervals with strengths of PE and FDI

Identification region	Public expenditure	FDI
Overall region	(0,1.690) 5.400 %	(1.920, 6.370) 12.00 %
Eastern region	(0,1.940) 5.000 %	(2.360, 6.390) 11.00 %
Middle region	(0,1.480) 6.000 %	(1.610,4.710) 8.000 %

identification is the highest at the interval in Fig. 2.1. And the effective area of FDI locates at interval (1.920, 6.370) of standardized GDP per capita in China. Second, the effective investment areas of public expenditure and FDI locate at interval (0, 1.940) and (2.360, 6.390) of standard GDP per capita in eastern region respectively. Third, as far as middle region is concerned, the effective investment area of public expenditure locates at (0, 1.480) and the one of FDI locates at interval (1.610, 4.710) of standard GDP per capita.

On the other hand, the efficiency strengths of FDI and public expenditure with path identifications are clearly depicted in Table 2.3. First, the efficiency strength of PE on GDP per capita distribution is 5.400 % at interval (0, 1.690) and the efficiency strength of FDI is 12.00 at interval (1.920, 6.370) in China. Second, the efficiency strengths are 0.050 % and 0.060 of PE on GDP per capita distribution for eastern and middle region, respectively. Similarly, the efficiency strength are 11.00 and 8.000 % of FDI on GDP per capita distribution for eastern and middle region, respectively. The efficiency strength is obtained by integration at effective interval, for example,

$$\int_0^{1.690} \left(f_{\omega(PE)}(x) - f_{benchmark}(x)\right)dx = 5.000\,\%$$

and

$$\int_{1.920}^{6.370} \left(f_{\omega(PDI)}(x) - f_{benchmark}(x)\right)dx = 5.000\,\%$$

2.3.4 *Joint Identification of Allocation Efficiency*

Actually, FDI and public expenditure work together in complex economic system. The FDI path or public expenditure path identification alone might not be accurate in describing the situation. In this sense, the joint path identification of allocation efficiency of FDI and public expenditure is valuable. This section provides the joint identification of GDP per capita distribution to reveal the joint allocation efficiency of FDI and public expenditure.

Similar to the single FDI path identification, the joint path identification is obtained by introducing the joint weight of FDI and public expenditure in Model (2.6). Figures 2.5, 2.6, 2.7 present the joint path identifications of GDP per capita distribution in 2006 for China, eastern China and middle China respectively. The joint path identifications illuminate that the joint investment effects are the combination effects due to FDI and public expenditure.

The joint path identification in China illustrates that the curve of joint path identification locates in the middle of the ones with FDI path and with benchmark path at interval (2, 6). The joint path presents a split of the difference between the negative impact of public expenditure and positive impact of FDI on average level of the distribution in this area.

Table 2.3 gives the effective intervals and efficiency strengths of joint path. First, the effective investment area of joint path locates at interval (2.180, 6.290) of

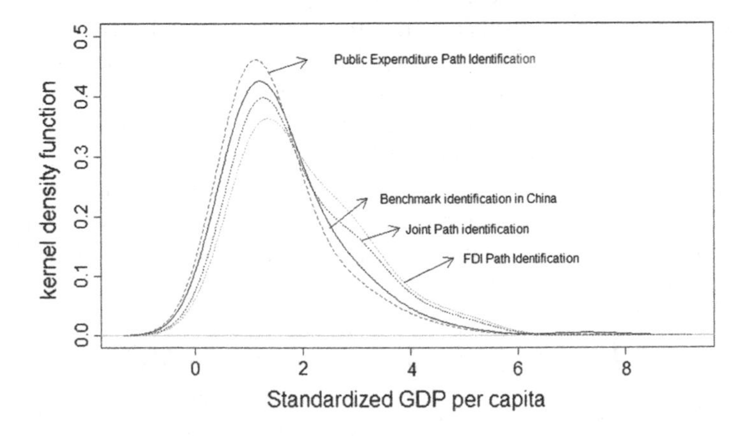

Fig. 2.5 Joint path identification of GDP per capita distribution in China in 2006

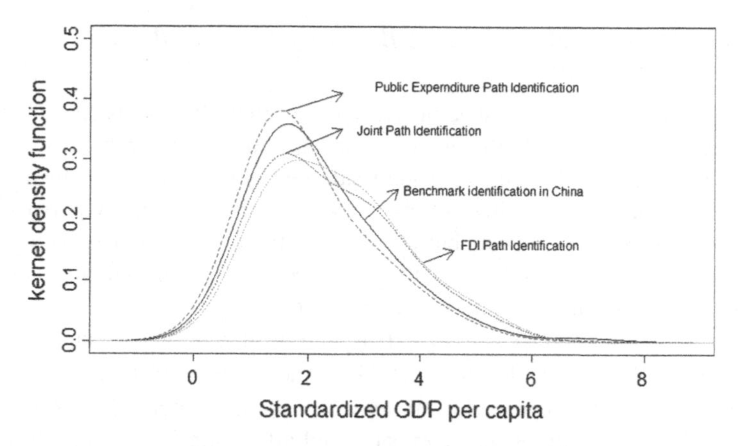

Fig. 2.6 Joint path identification of GDP per capita distribution in Eastern China in 2006

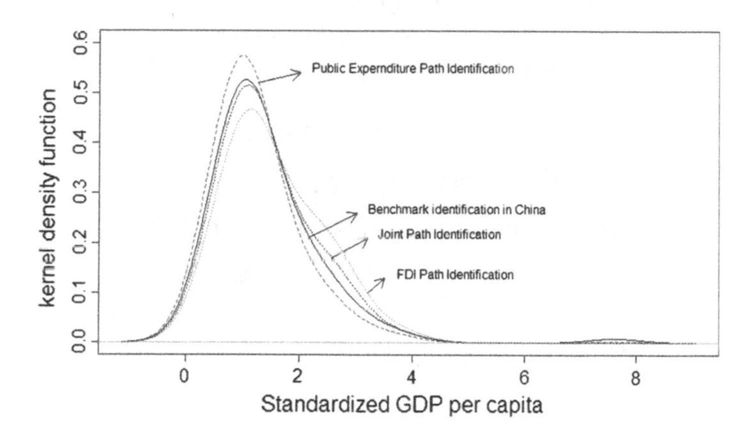

Fig. 2.7 Joint path identification of GDP per capita distribution in Middle China in 2006

Table 2.3 Effective intervals and efficiency strengths of joint path

Identification region	Effective intervals	Impact strengths (%)
Overall region	(2.180, 6.290)	8.700 (+)
Eastern region	(2.580, 6.290)	8.700 (+)
Middle region	(1.380, 4.660)	3.600 (+)

standardized GDP per capita in China. The effective areas of joint path locate at (2.580, 6.290) and (1.380, 4.660) in Eastern and Middle China respectively. Secondly, the efficiency strengths of joint path in China, Eastern and Middle China are 8.700, 8.700 and 3.600 %, respectively. The one in Eastern China is close to the level of whole China, which is probably because the amount of FDI and public expenditure in Eastern region accounts for the largest share in China.

2.4 Allocation Strategy with Efficiency-Oriented PE

An investment strategy for a specific enterprise generally selects the investment with efficiency-oriented with the objective of maximizing the profit. However, for a city (or region), the investment strategy adopted is not simply oriented by efficiency promotion; the initial investment strategy oriented by equity pursuit is also considered. The possible investment strategy for the city (or region) is to reverse the initial inefficient allocation to efficient allocation by introducing additional investment or reallocate the initial investment resource.

The identification indicates public expenditure is allocation inefficiency at cities where satisfy $f_\omega(x_j) - f_n(x_j) < 0$. Oriented by the investment philosophy of efficiency promotion, the investment strategy for public expenditure is to realize allocation efficiency at cities where experienced allocation inefficiency. The study in this section will measure the allocation strategies with efficiency-oriented public expenditures among different regions of China.

The theoretical optimization of the re-allocation of public expenditure p_j should change inefficient allocation $f_\omega(x_j) - f_n(x_j) < 0$ to

$$f_\omega(x_j) - f_n(x_j) = 0 \quad \text{for some city } j \tag{2.26}$$

where $f_n(x) = \frac{1}{n}\sum_{i=1}^{n} h^{-1}K\left(\frac{x-X_i}{h}\right)$, and $f_\omega(x) = \sum_{j=1}^{n} \omega_j h^{-1}K\left(\frac{x-X_i}{h}\right)$.

However, it is hard to obtain the unique solution of Eq. (2.30).

Fortunately, since integral of probability density satisfies $\int_{-\infty}^{\infty} f_\omega(x)dx = 1$ and $\int_{-\infty}^{\infty} f_n(x)dx = 1$, theoretically, it is easy to have $\int_{-\infty}^{\infty} (f_\omega(x) - f_n(x))dx = 0$

Then

$$\int_{D} (f_\omega(x) - f_n(x))dx = \int_{D} (f_n(x) - f_\omega(x))dx$$

where $\bar{D}$ is inefficient interval, D is efficient interval.

Thus, the optimal reallocation strategy of public expenditure is to replace the relationship in Eq. (2.26) by the average integral in Eq. (2.27) below.

Given $\varepsilon > 0$, such as inefficient strength is less than ε, i.e.,

$$\int_{-\infty}^{\infty} (f_n(x) - f_\omega(x))I(f_n(x) - f_\omega(x) > 0)dx = \int_{D} (f_n(x) - f_\omega(x)) \leq \varepsilon \tag{2.27}$$

Where $I(u \geq 0) = 1$, if $u \geq 0$, otherwise, $I(u \geq 0) = 0$, if $u < 0$.

2.4.1 Efficiency-Oriented Strategy of PE in Eastern Region

1. Reallocation of PE

The reallocation is implemented at both the inefficient and efficient cities in Eastern region. More specifically, it increases the public expenditure in cities at allocation inefficient interval and decreases the public expenditure in cities at allocation efficient interval; no extra public expenditure is added in the reallocation process. The realization of reallocation process illuminates the allocation in actual world.

The decision of public expenditure reallocation for cities in Eastern region is performed by following five steps.

First, the aim of public expenditure reallocation is to reduce the inefficiency strength to less than $\varepsilon = 2.000\,\%$. i.e.

$$\int_D (f_n(x) - f_\omega(x))dx = \int_D (f_\omega(x) - f_n(x))dx \leq 2.000\,\%$$

Second, the total amount of reallocated public expenditure is 11.14 billion Yuan, which is 1.180 % of original public expenditure in Eastern region. The total amount of reallocated public expenditure is

$$\alpha \sum_{k=1}^{n} pe_k I(f_n(x_k) - f_\omega(x_k) < 0) = 11.14 \text{ billion Yuan}$$

where $\alpha = 8.800\,\%$ is the proportion of the total efficiency allocated PE,. pe_k is the original amount of public expenditure in city k .

Third, the decreased public expenditure for Eastern city i is given by

$$p_{iE} = \frac{(f_n(x_i) - f_\omega(x_i))I(f_n(x_i) - f_\omega(x_i) < 0)}{\sum_{j=1}^{n} (f_n(x_i) - f_\omega(x_i))I(f_n(x_j) - f_\omega(x_j) < 0)} \times \alpha \sum_{k=1}^{n} pe_k I(f_n(x_k) - f_\omega(x_k) < 0) \tag{2.28a}$$

The increased public expenditure for Eastern city i is given by

$$p_{iI} = \frac{(f_n(x_i) - f_\omega(x_i))I(f_n(x_i) - f_\omega(x_i) \geq 0)}{\sum_{j=1}^{n} (f_n(x_i) - f_\omega(x_i))I(f_n(x_j) - f_\omega(x_j) \geq 0)} \times \alpha \sum_{k=1}^{n} pe_k I(f_n(x_k) - f_\omega(x_k) < 0) \tag{2.28b}$$

It illustrates re-allocation of public expenditure in Eastern cities. It is implemented according to the difference between benchmark and PE path models. The greater the difference is, the larger public expenditure will be.

Fourth, the new public expenditure for Eastern city i is obtained by

$$p_{1i} = pe_i - p_{iE} + p_{iI} \tag{2.29}$$

Fifth, the new public expenditure path weight is given by new path $\omega_i = l_i / \sum_{j=1}^{n} l_j \; l_i = p_{1i}/GDP_i$ in (2.19).

Five steps present the path-converged design of the new PE path after reallocation of public expenditure in Eastern cities.

2. Additional PE

The realization of additional PE only increases PE in ineffective cities and maintains the original PE in the effective cities.

With the aim of reducing the inefficiency strength to less than 2 %, the total additional public expenditure is 63.90 billion Yuan, which is 6.770 % of original public expenditure in Eastern region, and α is 50.50 % in this case. The increased public expenditure is given by Eq. (2.28b).

The new public expenditure in Eastern city i is obtained by

$$p_{1i} = p_i + p_{iI} \tag{2.30}$$

Both the realizations of reallocation and additional investment of public expenditure successfully reduce the inefficiency strength in Eastern region to 2 %. For reference, Fig. 2.8 presents the new PE path identifications of GDP per capita distribution in Eastern China.

Clearly, the strategy of additional PE is more effective than the strategy of reallocation PE according to the difference between PE path and benchmark identification in Fig. 2.8

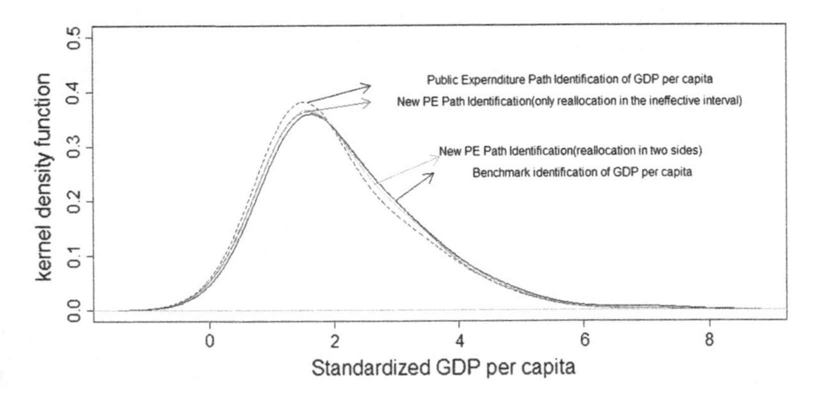

Fig. 2.8 New PE path identification of GDP per capita distribution in Eastern China in 2006

2.4.2 Efficiency-Oriented Strategy of PE in Middle Eastern Region

1. Reallocation of PE

First, the aim of public expenditure reallocation is to reduce the inefficiency strength to 2 %.

Second, the total amount of reallocated public expenditure is 12.72 billion Yuan, which is 4.730 % of original public expenditure in Middle region.

Third, the decreased and increased PE for middle city i is obtained by Eq. (2.28a) and (2.28b) respectively, where $\alpha = 11.40\,\%$.

2. Additional PE

With the aim of reducing the inefficiency strength to 2 %, the total additional public expenditure is 34.59 billion Yuan, which is 12.86 % of original public expenditure in Middle region, and α is 31.00 % in this case. The increased public expenditure is given by Eq. (2.28b).

Figure 2.9 presents the new PE path identifications of GDP per capita distribution in Middle China.

2.4.3 Efficiency-Oriented Strategy of PE in Overall Region

1. Reallocation of PE

First, the aim of public expenditure reallocation is to reduce the inefficiency strength to 2 %.

Second, the total amount of reallocated public expenditure is 27.83 billion Yuan, which is 2.020 % of original public expenditure in overall China.

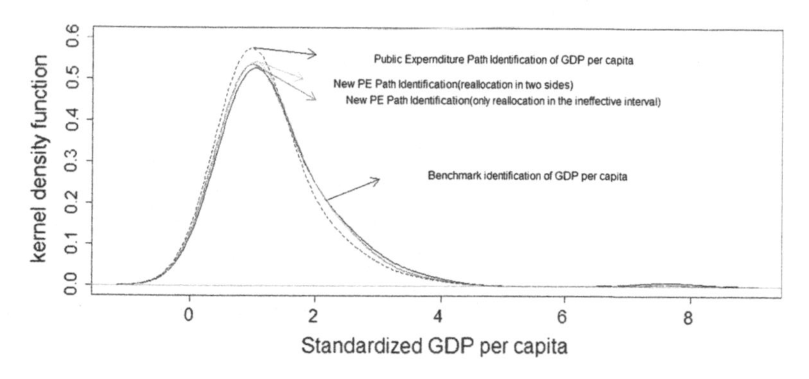

Fig. 2.9 New PE path identification of GDP per capita distribution in Middle China in 2006

Third, the decreased and increased PE for city i is obtained by Eqs. (2.28a) and (2.32b) respectively, where $\alpha = 8.300\,\%$.

2. Additional PE

With the aim of reducing the inefficiency strength to 2 %, the total additional public expenditure is 101.7 billion Yuan, which is 7.380 % of original public expenditure in whole China, and α is 30.50 % in this case. The increased public expenditure is given by Eq. (2.28b).

Figure 2.10 presents the new PE path identifications of GDP per capita distribution in China.

2.5 Allocation Strategy with Efficiency-Oriented FDI

2.5.1 Efficiency-Oriented Strategy of FDI in Eastern Region

1. Additional FDI

The realization of additional FDI only adds extra FDI in ineffective cities and maintains the original FDI in the effective cities.

With the aim of reducing the inefficiency strength to *less than 3 %*, the total amount of additional FDI is 1.978 billion Dollars, which is 3.450 % of original FDI in Eastern cities. α is 28.60 % in this case. The increased FDI for Eastern city i is given by

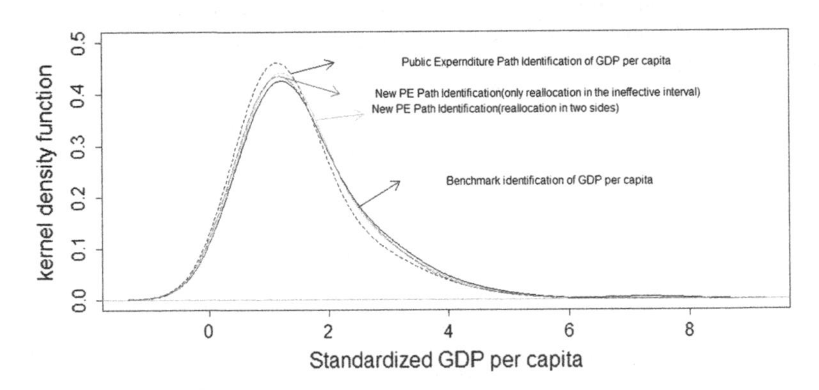

Fig. 2.10 New PE path identification of GDP per capita distribution in China in 2006

$$fdi_{iI} = \frac{(f_n(x_i) - f_\omega(x_i))I(f_n(x_i) - f_\omega(x_i) \geq 0)}{\sum_{j=1}^{n}(f_n(x_i) - f_\omega(x_i))I(f_n(x_j) - f_\omega(x_j) \geq 0)} \times \alpha \sum_{k=1}^{n} fdi_k I(f_n(x_k) - f_\omega(x_k) \geq 0) \tag{2.31}$$

The new FDI for Eastern city i is obtained by

$$fdi_{1i} = fdi_i + fdi_{iI} \tag{2.32}$$

The realizations of additional investment of FDI successfully reduce the inefficiency strength in Eastern region to 3 %. For reference, Fig. 2.11 presents the new PE path identifications of GDP per capita distribution in Eastern China.

2.5.2 Efficiency-Oriented Strategy of FDI in Eastern Region

With the aim of reducing the inefficiency strength to 3 %, the total amount of additional FDI is 663.950 million Dollars, which is 4.680 % of original FDI in Middle cities. The additional FDI for middle city i is given by Eq. (2.32), where $\alpha = 24.20\,\%$. Figure 2.12 presents the new PE path identifications of GDP per capita distribution in Middle China.

2.5.3 Efficiency-Oriented Strategy of FDI in Overall China

With the aim of reducing the inefficiency strength to 3 %, the total amount of additional FDI is 3.043 billion Dollars, which is 4.170 % of original FDI in overall cities. The additional FDI for middle city i is given by Eq. (2.32), where

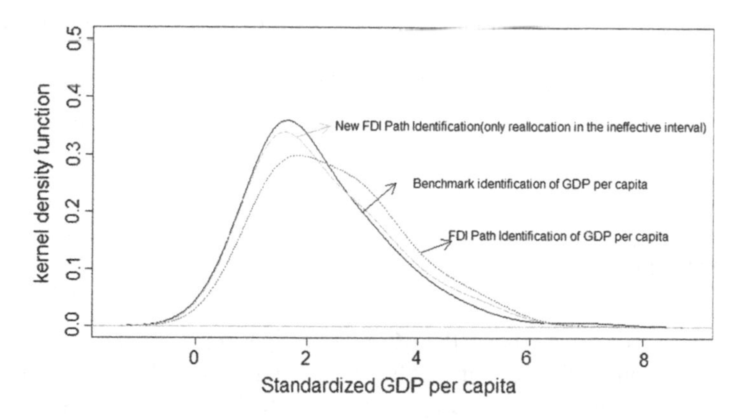

Fig. 2.11 New FDI path identification of GDP per capita distribution in Eastern China in 2006

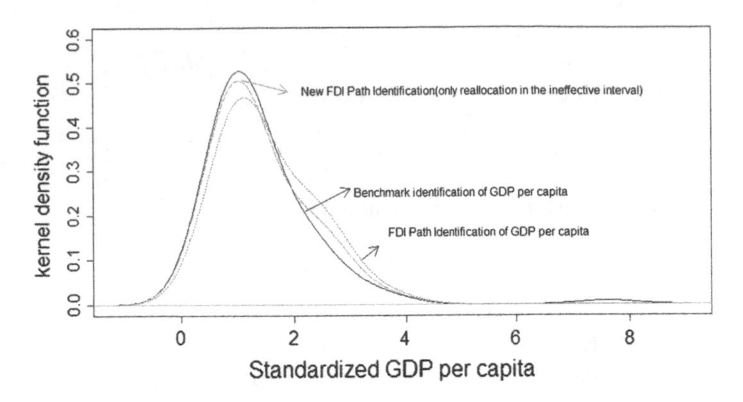

Fig. 2.12 New FDI path identification of GDP per capita distribution in Middle China in 2006

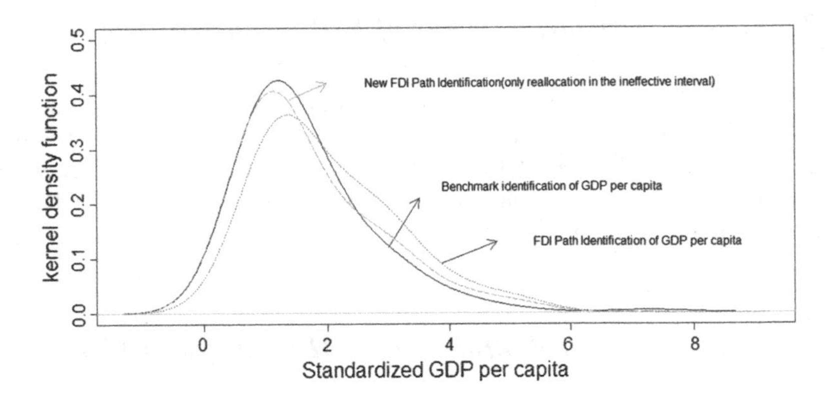

Fig. 2.13 New FDI path identification of GDP per capita distribution in China in 2006

$\alpha = 30.00\,\%$. Figure 2.13 presents the new PE path identifications of GDP per capita distribution in overall China.

With the aim to reduce the inefficiency strengths of public expenditure and FDI to 2 and 3 %, the realizations of reallocation and additional investment of public expenditure are presented. The realizations oriented by objective of promoting allocation efficiency hint the possible strategy for the economic stimulus package in China.

2.6 Conclusion

The study in this chapter sets up Path-Converged Design approach to identify the existing allocation efficiency of public expenditure and FDI among different regions in China. Furthermore, the investment strategies are presented to promote efficiency at areas where suffers from allocation inefficiency. Compared with previous studies, the new approach allows us to identify allocation efficiency from a perspective of dynamic distribution rather than the one of average impact parameters.

More specifically, the study first identifies the positive and negative impacts of PE and FDI on GDP per capita, secondly identifies the effective intervals with strengths of their allocations, and thirdly presents joint identification of allocation efficiency of PE and FDI, and finally realizes the optimal allocation of PE and FDI with changes from inefficient cities to efficient ones and changes from negative impacts to positive impacts in region due to PE and FDI.

The empirical results of path identifications and realizations of allocation efficiency for FDI and public expenditure on GDP per capita distribution in China and regions can be concluded as below.

First, the path identification identifies the production efficiencies of FDI and public expenditure. The single path identification illustrates FDI has positive impact and public expenditure has negative impact on GDP per capita, and the positive impact of FDI is greater than the negative impact of public expenditure in all regions, which is validated by the joint path identification that the production efficiency of joint path are in the middle of the effects of public expenditure path and the ones of FDI path.

Second, the identification of allocation efficiency is presented by the effective investment intervals and strengths. The public expenditure is more effective when invested into cities whose GDP per capita are less than average level of China; while FDI is more effective when invested into the cities whose GDP per capita are larger than average level of China. The situation may be probably interpreted by the difference of their policy directions. The public expenditure puts emphasis on equity while FDI focuses on the efficiency of production.

Third, the investment strategies to realize allocation efficiency with reallocation of public expenditure are implemented by increasing PE in cities at inefficient interval and decreasing PE in cities at efficient interval. The investment strategies to realize allocation efficiency with additional investment of PE and FDI pursue the increase of PE and FDI in ineffective cities and maintain the original PE and FDI in the effective cities. To successfully reduce the inefficiency strengths of PE to 2 %, the amounts of reallocated public expenditure are 11.14, 12.72, and 27.83 billion Yuan and the amounts of additional investment in PE are 63.90, 34.59, and 101.7 billion Yuan for Eastern, Middle and Overall China respectively. To successfully reduce the inefficiency strengths of FDI to 3 %, the amounts of additional investment in FDI are 1.978 billion, 663.9 million and 3.043 billion dollars for Eastern, Middle and Overall China respectively. The investment strategies oriented by allocation efficiency promotion also imply the possible strategy for the economic stimulus package in current China.

It is worthy to mention that new approach allows us to identify allocation efficiency from a perspective of distribution, rather than a perspective of parameters of production efficiency. Besides, the new approach presented in this chapter is capable of identifying further interior relationships between input and output of production efficiency, such as the relationships between exchange rates, unemployment rates and GDP.

References

Alfaro, L., Chanda, A., Kalemli-Ozcan, S., & Sayek, S. (2004). FDI and economic growth: the role of local financial markets. *Journal of International Economics, 64*(1), 89–112.

Balasubramanyam, V. N., Salisu, M., & Dapsoford, D. (1996). Foreign direct investment and growth in EP and IS countries. *Economic Journal, 106*(434), 92–104.

Blomstrom, M., Lipsey, R., Zejan, M. (1994). What explains the growth in developing countries? NBER Discussion paper, no. 1924.

Borensztein, E., De Gregorio, J., & Lee, J.-W. (1998). How does foreign direct investment affect economic growth? *Journal of International Economics, 45*(1), 115–134.

Goss, E., Wingender, JR., Torau, J. M. (2007). The contribution of foreign capital to U.S. productivity growth. *The Quarterly Review of Economics and Finance*, *47*(3), 383–396.

Kawai, H. (1994). International comparative analysis of economic growth: Trade liberation and productivity. *Developing Economies, 17*(4), 372–397.

Kim, T., & Lee, S. (2005). Kernel density estimator for strong mixing processes. *Journal of Statistical Planning and Inference, 133*(2), 273–284.

Lucas, R. E. (1988). On the mechanics of economic development. *Journal of Monetary Economics, 22*(3), 3–42.

Luo, C. (2007). FDI, domestic capital and economic growth: Evidence from panel data at China's provincial level. *Frontiers of Economics in China, 2*(1), 92–113.

Romer, P. M. (1986). Increasing returns and long-run growth. *Journal of Political Economy, 94*(5), 1002–1037.

Solow, R. M. (1957). Technical change and the aggregate production function. *Review of Economics and Statistics, 39*(3), 312–320.

Silverman, B.W. (1986). Density Estimation for Statistics and Data Analysis. Chapman & Hall, New York.

Stone, C. J. (1984). An asymptotically optimal window selection rule for kernel density estimates. *The Annals of Statistics, 12*, 1285–1297.

Xu, B., & Watada, J. (2008c). The path identification of FDI and financial expenditure in China, proceeding of fifth meeting international symposium on management engineering, Japan.

Chapter 3
Changes in Population Agglomeration Efficiency in Urban Planning

Most experts hold that contemporary China is in the transition from planned economy to market economy. Speaking of the planned economy, the most characterized phenomenon is the obligatory census register system, which aims to restrict internal population migration between cities and regions. The obligatory census register system does have positive roles on the management during the early period of new China. However, the obligatory census register system has greatly hindered the free migration of labor factors and urbanization process. While the 18th National Congress of the Communist Party of China emphasized that urbanization process is the main driving force for economic growth. Therefore, the reallocation of population resource through the channel of urban planning is of crucial importance to realize allocation efficiency in China.

This chapter continues the allocation efficiency framework in Chap. 2. The difference lies in the fact that Chap. 2 aims to identify and measure allocation efficiency of public expenditure and FDI, taken as additional investment factor in growth system, while Chap. 3 aims to identify and measure the allocation efficiency of population migration, taken as the labor input in growth system.

Specifically, this chapter develops identification based on nonparametric kernel density estimation with path-converged design to first identify the population agglomeration efficiency from both regional and urban perspectives, and secondly measure the population agglomeration efficiency with proposed decisions of optimal population migration in urban planning in China.

3.1 Introduction

Previous studies on urban planning mostly focus on population migration. Johnson (2004) estimated about 39.70 million immigrations and 16.300 million emigrations in Chinese provinces during 2000. Yang (2003) studied the scale and intensity of Chinese population migration and found that about 20 million rural people immigrated into the cities during 1955–2000. Yu and Zhang (2006) employed related methods of direction and distance in spatial statistics to study the

B. Xu et al., *Changes in Production Efficiency in China*,
DOI: 10.1007/978-1-4614-7720-4_3,

population immigration of Beijing, Tianjin and Tang metropolis area, Yangtze Delta metropolis area and Pearl River Delta metropolis area from the perspectives of immigration direction, immigration distance and immigration factors. Wang and Huang (2005) discovered that provincial immigration was an irreplaceable factor for pushing the development in Eastern China during 1995–2000.

In addition to investigations of population migration, studies tend to explore whether urbanization lags behind economic growth. Cai (2006) contends that urbanization progress tends to bring about production agglomeration rather than population agglomeration; i.e., the system itself partly causes urbanization development to lag behind economic growth. The current crisis that Chinese urbanization development faces is fundamentally a systematic supply crisis (Chen 2006). Studying provincial panel data during 1987–2004, Lu and Chen (2004) showed that the widening income gap between urban and rural areas was related to policies favoring urban areas of the regional government.

In contrast to previous focuses on how to control free migration and whether urbanization lags behind economic growth in China, the study first aims to identify existing population agglomeration for small, medium, and large cities from both regional and urban perspectives and to evaluate the efficiency of existing population agglomeration in urban planning. Furthermore, the study aims to measure the population agglomeration efficiency with proposed optimal strategy of population agglomeration in urban planning.

Inspired by Henderson's (2003) investigation of how the significant deviation from any optimal degrees of urban concentration or rates of urbanization causes economic losses in terms of the maximization of productivity growth, this study develops a path-converged design based on a nonparametric kernel density distribution approach to identify of population agglomeration efficiency and to propose decision making for population agglomeration with the aim of eliminating inefficiency in China. The study in this chapter is the further study based on the existed investigations (Xu and Watada 2006, 2007a, 2008a, b; Xu et al. 2008).

Section 3.2 briefly illustrates identification approach of population agglomeration efficiency. Section 3.3 identifies the population agglomeration efficiency from the perspectives of region, urban and city size. Section 3.4 measures the population agglomeration efficiency with proposed population migration strategy in urban planning. Section 3.5 finally concludes.

3.2 Path-Converged Design

Path-converged design based on nonparametric econometrics is developed in this study rather than the typical linear regression, Granger cointegration, and ARCH models. China's population policy has caused a great disparity between the actual urban population and the registered urban population, according to household register system. For example, the floating population in Beijing is 3.573 million; 23 % of those people were permanent 2005, and 38.80 % of those people had worked and lived in Beijing for more than 5 years. They are essentially "Beijing residents". Moreover, the statistical standard has experienced several changes and adjustments, so the observation of population data tends to vary for many reasons. The urbanization level of observed data from the household register system heavily deviates from the actual urbanization level. Even if cross-sectional or time series data are accepted by unit root tests, the analysis of models such as linear regression and ARCH are doubtful because the data distribution itself presents a fat-tail feature, and higher order moment does not exist.

The identification approach of population agglomeration originates from the change of urbanization structure with following three processes, which is under the same framework in Chap. 2. First, suppose initial population allocation is equilibrium of urbanization structure. Second, when new populations immigrate into the initial equilibrium, the initial structure will change to a dynamic disequilibrium structure of population allocation. Finally, the disequilibrium structure will come to re-equilibrium after some time to adjust the disequilibrium into equilibrium state of resource allocation.

3.2.1 Benchmark Model

The estimation of the underlying structure of urbanization is given by the kernel density estimation:

$$S_1 : f_n(x, X) = \frac{1}{n} \sum_{i=1}^{n} h^{-1} K\left(\frac{x - X_i}{h}\right) \tag{3.1}$$

where $K(\bullet)$ stands for the kernel function and h the bandwidth.

Equation (3.1) presents an estimation of urbanization level distribution density with combination of information in all prefecture-level cities. It gives the equal weight $1/n$ at each observation X_i. That is, arithmetical average weighted kernel density estimation **models** *the equilibrium* of resource allocation on overall urbanization level.

3.2.2 Path Model

Because h is related to corresponding weights in the study, conventional bandwidth selection methods irrelevant to ϖ_j and j are inappropriate. This study tries to obtain the bandwidth through the following steps.

The choice of bandwidth h determines the amount of smoothing needed to estimate kernel density. Inspired by the application of Maria and Roberto (2004), we define the number of modes of weighted density $f_\omega(x)$ as:

$$M(f_{\omega}) = \#\{y \in \Re_{+} : f'_{\omega} = 0 \; and \; f''_{\omega}(y) < 0\}$$

Null hypothesis H_0: *the underlying density* f_{ω} *has m modes* $(M(f_{\omega}) \leq m)$;

$$H_1 : f_{\omega} \text{ has more than m models } (M(f_{\omega}) > m)$$

for $m = 1, \ldots, M$, where $\hat{f}_{\omega h}$ is the weighted kernel density estimate and h the overall bandwidth.

Bootstrap multimodality test is based on the notions of critical smoothing and critical bandwidth. The mth critical bandwidth $\hat{h}_{m,crit}$ is defined by:

$$\hat{h}_{m,crit} = \inf\{h : M(\hat{f}_{\omega h}) \leq m\}$$

when $h = \hat{h}_{m,crit}$, $\hat{f}_{\omega h}$ will display m modes with a shoulder. When h further decreases, an additional $(m+1)$th mode will appear in the place of the shoulder.

Thus, this study attempts to select h through following process:

Step 1. Apply the Cross Validation method to the benchmark model (3.1) on the rule of *IMSE*, and then select the initial pilot bandwidth h_1;
Step 2. For $m = 1, \ldots, M$ of weighted density function $f_{\varpi h}(x)$, obtain each $\hat{h}_{m,crit}$.
Step 3. Bootstrap test on the number of modes m, for $m = 1, \ldots, M$.
(a) As suggested by Silverman (1981), the bootstrap data x^j are generated by:

$$x_i^j = \bar{y}^j + \left(1 + \hat{h}^2_{m,crit}/\hat{\sigma}^2\right)^{-1/2}\left(y_i^j - \bar{y}^j + \hat{h}_{m,crit}\varepsilon\right) \tag{3.2}$$

where y_i^j are randomly drawn (with replacement) from the original sample, $\bar{y}^j$ is its mean, $\hat{\sigma}^2$ is its variance, and ε is an *i.i.d* $N(0,1)$ variable for $j = 1, \ldots, b$, $i = 1, \ldots, n$;

(b) For $j = 1, \ldots, b$, apply the weighted kernel method to obtain bootstrap estimate $\hat{f}^j_{\omega, h_{m,crit}}$ based on the critical bandwidth $\hat{h}_{m,crit}$ that is computed from the original data. For each bootstrap sample, calculate the corresponding number of modes $M\left(\hat{f}^j_{\varpi, h_{m,crit}}\right)$ and the m-th critical bandwidth $\hat{h}^j_{m,crit}$.
(c) Compute the estimate of the achieved significance level (ASL) or p value:

$$\widehat{ASL}_m = p_m = \frac{1}{b}\sum_{j=1}^{b} \#\left\{M(\hat{f}^j_{h_m,crit}) > m\right\} \tag{3.3}$$

(d) Do not reject the null of m modes in the underlying density if $\widehat{ASL}_m$ or p_m is sufficiently large.

The test is performed by simultaneously computing $\widehat{ASL}_m$ for all $m = 1, \ldots, M$, where M is a predetermined number. For the underlying density f_ω, we select an estimate $\hat{m}$ as the critical number of modes if $\widehat{ASL}_m$ shows the largest value among all $\widehat{ASL}_m$ for $m = 1, \ldots, M$.

With the bandwidth h and kernel function $K(\bullet)$, the benchmark model $f_h(x)$, regional population path distribution $f_{\omega 1,h}(x)$ and urban population path $f_{\omega 2,h}(x)$ are obtained.

$$f_{\omega n}(x) = \sum_{j=1}^{n} \omega_j h^{-1} K\left(\frac{x - X_i}{h}\right) \tag{3.4}$$

where $K(\bullet)$ is a kernel function, and h is the bandwidth, $f_{\omega 1,h}(x)$ and $f_{\varpi 2,h}(x)$ are the regional population factor path and the urban population factor path of urbanization level, respectively.

3.3 Population Agglomeration Efficiency Identification

3.3.1 Data Description

The source of data for this study is the Chinese City Statistical Yearbook at www.bjinfobank.com. The variables included in the study are: (1) year-end total population in urban area (2) year-end non-agriculture population in urban area, (3) year-end total population in entire city, and (4) year-end non-agriculture population in entire city, (5) GDP in urban area and (6) GDP in entire city of prefecture-level cities in 2002, 2004, and 2006. The prefecture-level cities in this study refer to all prefecture-level cities, cities at sub-provincial levels, and municipalities directly managed by the central government in China. According to the classifications in China, the entire city of the prefecture-level city is the administrative area under the city's jurisdiction, including the countryside or numerous counties and county-level cities, and which refers to the regional area in this study. The urban municipal district refers to the urban city and its suburbs, which are taken as an urban area in this study.

Two definitions are widely used to describe the development of the urbanization level: the proportion of urban population and the proportion of non-agricultural population. The first definition is usually used for the analysis of single urbanization development levels in China. In this study, *non-agriculture population data for prefecture-level cities, rather than urban population data, are available*. Moreover, this study seeks to obtain the overall urbanization development level from the urbanization development levels of nearly 300 prefecture-level cities, focusing on the relativity rather than the absoluteness of the data. In this sense, the study is different from the single index of urbanization level.

Accordingly, this study takes the urbanization level to be the proportion of non-agriculture population. This embodies the shift of rural surplus labor to the city and also expresses the labor receiving capability of the city. The statistical data illustrate that the urbanization levels by the second definition in 2002, 2004, and 2006 are 27.89, 30.81, and 36.26 %, respectively, which are lower than the urbanization level defined by the proportion of urban population (Li and Jin 2005).

The reason for employing urbanization development in prefecture-level cities to represent the urbanization level is that China's growth was sustained mainly by these cities in the 1990s. The total GDP of over 290 prefecture-level cities accounted for more than 50 % of GDP in 2000, and their average growth rates are mostly over 12 %. China might have failed to deal with deflation in the 1990s if not for double-digit growth of the prefecture-level cities.

The research aims to identify the existing population agglomeration and its efficiency in urban planning with the path-converged design in China. It is pursued from the following two perspectives:

The study seeks to identify population agglomeration and its efficiency in urban planning in China by using a path-converged design from the following two perspectives:

1. the perspective of city size, which separates nearly 300 prefecture-level cities in China into large, middle-sized, and small cities according to an urban population scale, which is different from the usual single urbanization level;
2. the perspective of spatial location, which separates city regions into urban and regional areas according to the scale of the city region.

3.3.2 Urbanization Level: Benchmark Model

Figure 3.1 shows the benchmark distribution of urbanization levels in prefecture-level cities during 2002, 2004, and 2006 and reflects the agglomeration of cities. Table 3.1 displays detailed information regarding the distributions in each year. $f_h(x)_{\max}$ is the maximum value of function $f_h(x)$, S_1, S_2, and S_3 are the integral values of each distribution at the intervals (0, 0.35], (0.35, 0.7], and (0.7, 1], satisfying $S_1 \int_0^{0.35} f(x)dx$, $S_2 \int_{0.35}^{0.7} f(x)dx$, $S_3 \int_{0.7}^{1} f(x)dx$. The intervals of (0, 0.35], (0.35, 0.7], and (0.7, 1] illustrate low, medium, and high urbanization levels. *EX* is the expected value of overall urbanization development level in China with each distribution, satisfying $EX = \int_0^{\infty} x f_h(x)dx$—that is, the overall average urbanization level in each year. The bottom row in Table 3.1 presents the average growth rate of each index during 2002–2006 (if negative, it is declining). The average growth rate of the expected value is 1.790 %, obtained by solving $0.3150(1+x)^4 = 0.3380$.

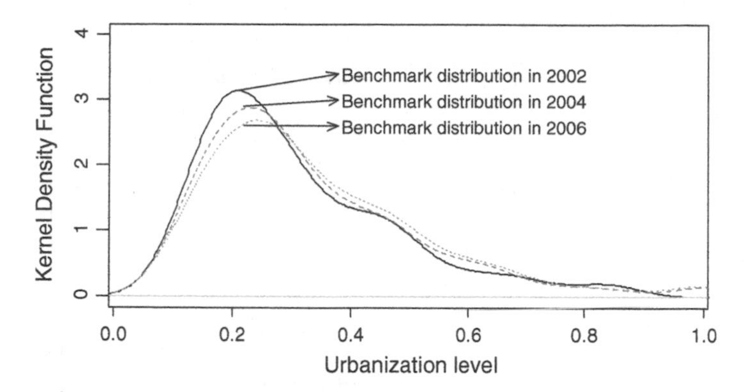

Fig. 3.1 Benchmark distribution of urbanization in prefecture-level cities in 2002, 2004 and 2006

Table 3.1 The information regarding benchmark distributions in 2002, 2004, 2006

Years	$\widehat{f}_h(x)_{\max}$	*EX*	S_1	S_2	S_3
2002	3.150	0.3150	0.6650	0.2920	0.04260
2004	2.880	0.3250	0.6290	0.3190	0.04280
2006	2.680	0.3380	0.5950	0.3460	0.04870
Growth Rate	−3.880 %	1.790 %	−2.740 %	3.300 %	3.410 %

First, Table 3.1 and Fig. 3.1 illustrate that most cities are located at the low urbanization level interval (0, 0.35), since the average urbanization levels are less than 0.35 and the integral of the benchmark distributions at (0, 0.35) takes up at least 60 % for all years.

Second, the urbanization level gradually grows over time. On one hand, the average growth rate of urbanization during 2002–2006 is 1.790 % for prefecture-level cities in China. On the other hand, the integral area at higher urbanization increases gradually; in particular, the integral area at interval (0.7, 1) grows at 3.410 % every year, while the integral area at interval (0, 0.35) decreases gradually with a declining rate of 2.740 % every year. Similarly, the highest point of the function slips every year at an annual rate of 3.880 %.

3.3.3 Population Agglomeration: Path Model

In order to identify population agglomeration modes from both regional and urban perspectives, the following section presents the regional population path and urban population path distributions for analysis.

Figure 3.2 and Table 3.2 list information about the regional population path distributions, reflecting the regional population agglomeration modes of cities.

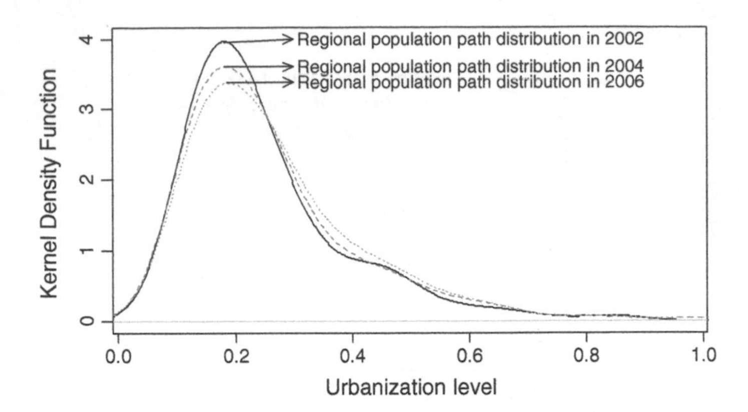

Fig. 3.2 Regional population path distributions of urbanization in prefecture-level cities in 2002, 2004 and 2006

Table 3.2 Regional population path distributions in 2002, 2004, 2006

Years	$\widehat{f}_h(x)_{\max}$	EX_1	S_1	S_2	S_3
2002	3.970	0.2530	0.8010	0.1800	0.01640
2004	3.600	0.2600	0.7820	0.1960	0.01780
2006	3.370	0.2700	0.7550	0.2210	0.01920
Growth Rate	−4.010 %	1.660 %	−1.480 %	4.210 %	4.020 %

The average growth rate of the average regional urbanization is 1.660 % during 2002–2006. Similar to the situation in the benchmark, the integral area at higher urbanization levels increases gradually; in particular, the integral area at interval (0.700, 1) grows at 4.020 % every year, while the integral area at interval (0, 0.350) decreases gradually at a rate of 1.480 % every year. Again, the maximum value of the regional population path distributions slips yearly at an annual rate of 4.010 %.

Figure 3.3 and Table 3.3 illustrate the details of urban population path distributions, giving detailed information on urban population agglomeration modes of cities. The average growth rate of average urban urbanization is 1.870 % during 2002–2006. The integral area at higher urbanization increases gradually; in particular, the integral area at interval (0.7, 1) grows at 2.740 % every year, while the integral area at interval (0, 0.35) decreases gradually at a rate of 1.330 % every year. Again, the maximum value of the urban population path distributions decreases yearly at an annual rate of 3.840 %.

In sum, two conclusions are presented regarding the population agglomeration modes of both regional and urban perspectives.

First, both the regional and urban populations agglomerate to cities with urbanization lower than 0.35 because the integrals of both the regional population path and urban population path distributions at (0, 0.35) take up at least 75 % for all years, which presents the inefficiency of population agglomeration.

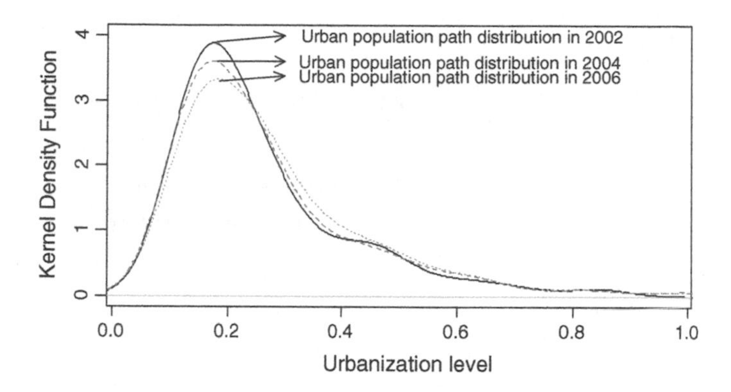

Fig. 3.3 Urban population path distributions of urbanization in prefecture-level cities in 2002, 2004 and 2006

Table 3.3 Urban population path distributions in 2002, 2004, 2006

Years	$\hat{f}_h(x)_{\max}$	EX_2	S_1	S_2	S_3
2002	3.890	0.2600	0.7830	0.1920	0.02360
2004	3.610	0.2650	0.7690	0.2000	0.02520
2006	3.330	0.2760	0.7420	0.2260	0.02630
Growth Rate	−3.840 %	1.520 %	−1.330 %	4.160 %	2.740 %

Second, the agglomeration of both regional and urban populations gradually shrinks over time because the integral areas at (0, 0.35) witness a declining trend for both path distributions.

3.3.4 Agglomeration Inefficiency in Region and Urban Area

This section aims to identify the population agglomeration efficiency from both perspectives of region area or urban area through the channel of urbanization level.

Figures 3.4, 3.5, 3.6 illustrate the comparisons of distributions among benchmark, regional population path and urban population path distributions in 2002, 2004 and 2006. Table 3.4 lists the average urbanization levels of three distributions in 2002, 2004 and 2006.

In Table 3.4, EX, EX_1 and EX_2 represent the expected values of benchmark, regional population path, and urban population path distributions, respectively.

All the figures show the same trend: both the regional and urban population path distributions drift *leftward* based on their corresponding benchmark distribution, which hints the inefficiency of population agglomeration. The regional population path distributions experience a stronger shift compared with the urban population path distributions. This trend implies that both regional and urban

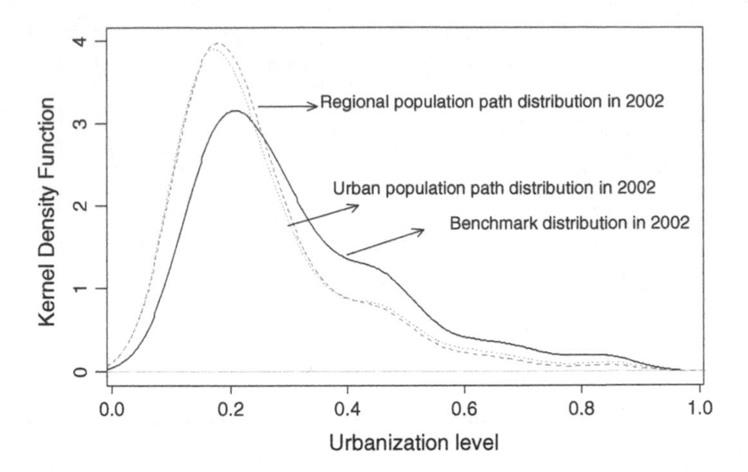

Fig. 3.4 Comparison in 2002

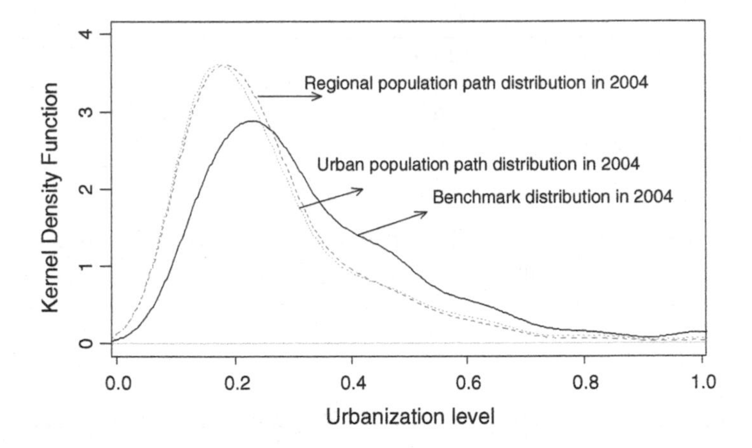

Fig. 3.5 Comparison in 2004

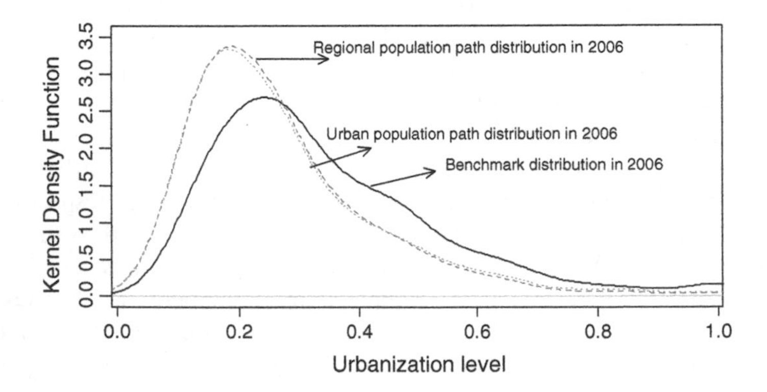

Fig. 3.6 Comparison in 2006

Table 3.4 The average urbanization levels of the distributions of benchmark, regional population and urban population path in 2002, 2004 and 2006

Years	EX	EX_1	$EX_1 - EX$	EX_2	$EX_2 - EX$
2002	0.3150	0.2530	−0.06170	0.2600	−0.05460
2004	0.3250	0.2600	−0.06500	0.2650	−0.06010
2006	0.3380	0.2700	−0.06760	0.2760	−0.06140
Average	0.3260	0.2610	−0.06480	0.2670	−0.05870

population path distributions reduce the overall level of urbanization, which is also observed from the figures in Table 3.4.

The expected values of the benchmark distributions are highest, and the expected values of regional population path distributions are the lowest. The regional population path and urban population path reduce the benchmark expected values by 6.960 and 6.270 %, which are obtained by $EX_1 - EX$ and $EX_2 - EX$ respectively. This result implies that the population gathers into areas with urbanization levels lower than both average regional and urban urbanization levels. However, the strength of the decline of the regional population path distribution is slightly stronger than the strength of the decline of the urban population path distribution.

3.3.5 Population Agglomeration Inefficiency: City Size

The study shows in the previous section that the population mainly gathers into cities at lower urbanization levels from both regional and urban perspectives. In order to clarify whether the inefficiency of population agglomeration exists in small, medium and large cities, the research goes on to explore the agglomeration from city size perspective.

In this section the research separates all prefecture-level cities into three groups: small cities, with their total urban population less than 0.5 million, medium cities, with their total urban population range from 0.5 to 1 million, and large cities, with their total urban population larger than 1 million. The research gives the benchmark, regional population path and urban population path distributions for small cities, medium cities and large cities in 2000, 2002 and 2004, respectively, to identify the feature of population agglomeration. Figures 3.7, 3.8, 3.9, 3.10, 3.11, 3.12 present the three distributions for three city groups in 2004 and 2006, and Table 3.5 lists their corresponding expected values.

Compared with the benchmark, both regional and urban population path distributions in each city group in all years drift leftward, which is consistent with the situation of all prefecture-level cities. The regional population path distribution and urban population path distribution tends to be the same for small cities in both 2004 and 2006, however the urban population path distributions tend to shift

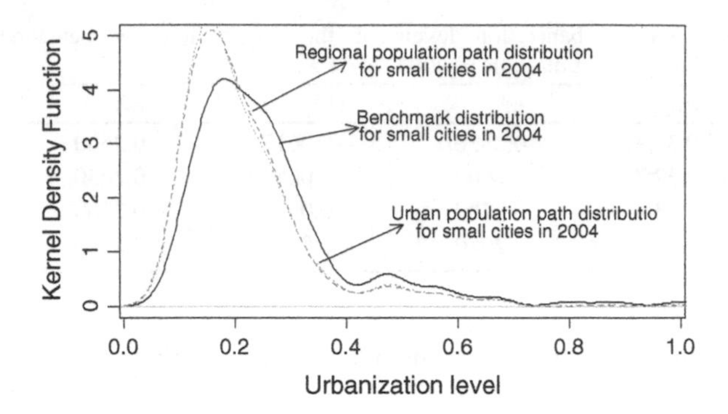

Fig. 3.7 Distributions of small cities in 2004

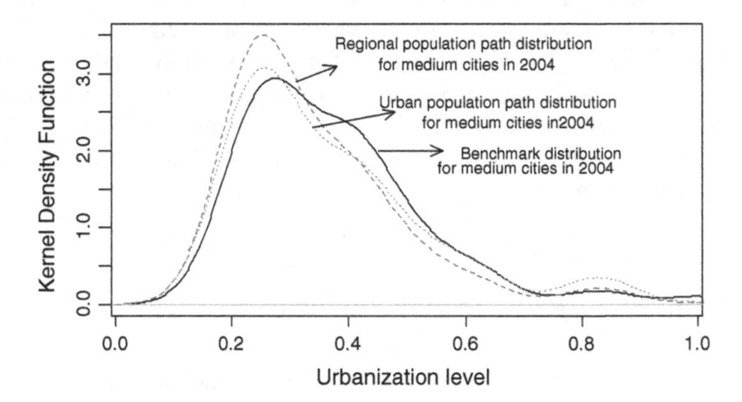

Fig. 3.8 Distributions of medium cities in 2004

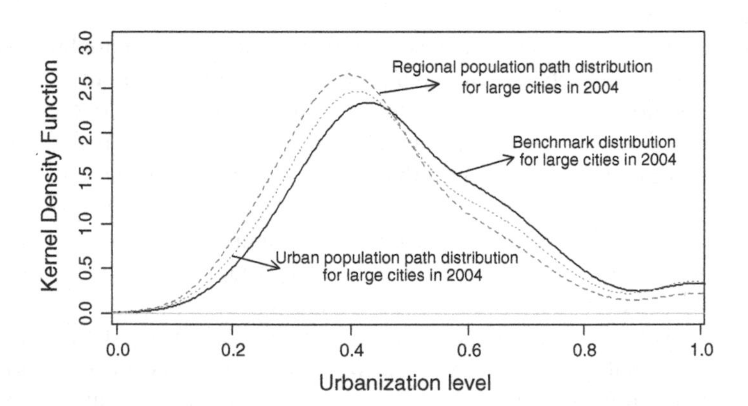

Fig. 3.9 Distributions of large cities in 2004

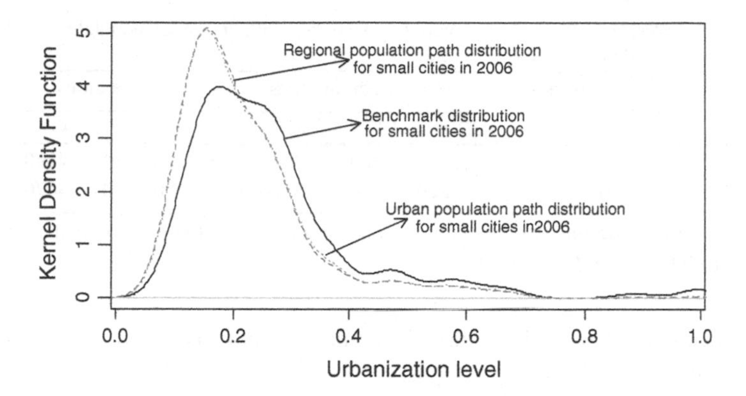

Fig. 3.10 Distributions of small cities in 2006

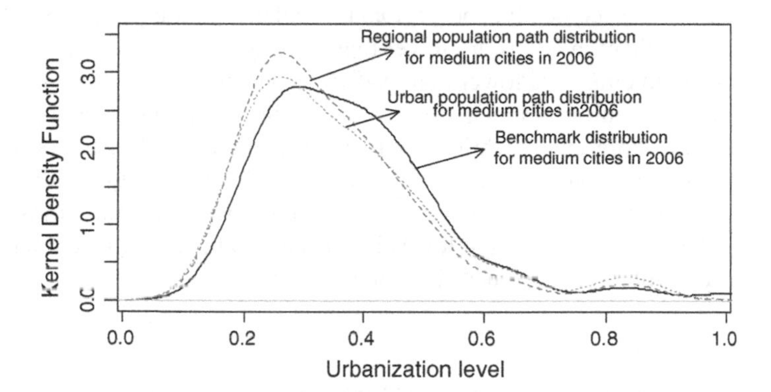

Fig. 3.11 Distributions of medium cities in 2006

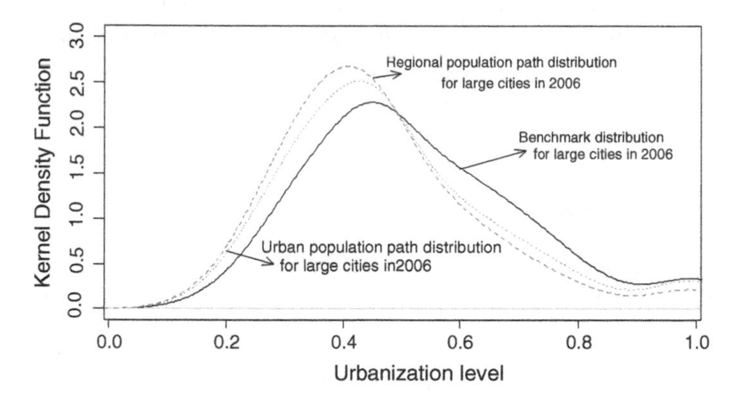

Fig. 3.12 Distributions of large cities in 2006

Table 3.5 The average urbanization levels of the benchmark, regional population and urban population path distributions for small cities

City category year	Small cities			Medium cities			Large cities		
	EX	EX_1	EX_2	EX	EX_1	EX_2	EX	EX_1	EX_2
2004	0.2550	0.2150	0.2160	0.3690	0.3420	0.3680	0.4830	0.4390	0.4630
2006	0.2600	0.2190	0.2220	0.3760	0.3480	0.3642	0.4990	0.4470	0.4640

EX, EX_1 and EX_2 denote the expected value of benchmark, regional population path, and urban population path distributions, respectively

rightward compared with the regional population path distributions for medium and large cities in both years.

The inefficiency of population agglomeration mode is identified in large, medium and small cities because all the population gathers into the cities whose urbanization levels are ***lower than*** the average level from both regional and urban perspectives. More specifically, the regional population mainly gathers into the cities whose urbanization levels are lower than 0.2143 for small cities, lower than 0.3341 for medium cities and lower than 0.4919 for large cities in 2006. The urban population agglomeration is much similar to the regional population agglomeration in small and large cities, while the urban population mainly gathers into the cities whose urbanization levels are lower than 0.2935 for medium cities.

With the aim to promote urbanization, population migration from cities at low urbanization levels to cities at high levels is expected to be implemented based on the identified population agglomeration modes.

3.4 Decision Making on Optimal Migration Strategy

Identification indicates population agglomeration inefficiency exists from both the perspectives of city size, regional and urban area. The decision making is to induce population migration from cities at low urbanization level to cities at high urbanization level for eliminating inefficiency. The theoretical optimization of migration population p_j for city j satisfies the equation as follows:

$$f_{\omega}(x_j) - f_n(x_j) = 0 \tag{3.5}$$

where $f_n(x) = \frac{1}{n}\sum_{i=1}^{n} h^{-1}K\left(\frac{x-X_i}{h}\right)$, and $f_{\omega}(x) = \sum_{j=1}^{n} \omega_j h^{-1}K\left(\frac{x-X_i}{h}\right)$.

However, it is hard to obtain the unique solution of Eq. (3.1).
Fortunately, it has

$$\int_0^1 (f_{\omega}(x) - f_n(x))dx = 0 \tag{3.6}$$

Because the integral of probability density $\int_0^1 f_\omega(x)dx = 1$ and $\int_0^1 f_n(x)dx = 1$

Furthermore, population migrates from the cities at lower urbanization level to the cities at higher level, there are some values l, such that

$$\int_0^1 (f_\omega(x) - f_n(x))dx = \int_0^1 (f_n(x) - f_\omega(x))dx \tag{3.7}$$

l is estimated by Eq. (3.6).

Thus, the optimal strategies for migration population are given in the sense of Eq. (3.6) for small, medium and large cities in 2006.

3.4.1 Strategy 1: Regional Migration for Small Cities

l is 0.214 in the decision making of regional population migration for small cities. More specifically, the decision making of population migration for small cities is carried on with four steps.

First, the strength of migration population is 14.64 percent.

$$\int_0^{0.214} (f_{\omega_1}(x) - f_n(x))dx = 0.1464$$

Second, the total number of migration population is

$$0.1464 \sum_{k=1}^{n} p_k I(u_k < 0.214)$$

where p_k is the regional population of small city k, and $\sum_{k=1}^{n} p_k I(u_k < 0.214)$ is the total regional population of cities whose urbanization level lower than 0.214. $I(u_k < 0.214 = 1)$, if $(u_k < 0.214)$, otherwise, $I(u_k < 0.214) = 0$, if $u_k \geq 0.214$.

Third, the emigration population for some small city i is given by

$$p_{iE} = \frac{[f_{\omega_1}(u_i) - f_n(u_i)]I(u_i < 0.214)}{\sum_{j=1}^{n} [f_{\omega_1}(u_j) - f_n(u_j)]I(u_j < 02.14)} \times 0.1464 \sum_{k=1}^{n} p_k I(u_k < 0.214)$$

The immigration population for some small city i is given by

$$p_{iI} = \frac{[f_n(u_i) - f_{\omega_1}(u_i)]I(u_i \geq 0.214)}{\sum_{j=1}^{n} [f_n(u_j) - f_{\omega_1}(u_j)]I(u_j \geq 02.14)} \times 0.1464 \sum_{k=1}^{n} p_k I(u_k < 0.214)$$

It illustrates that the population migration for any small city is implemented according to the difference of benchmark and original regional population path distributions. The greater the difference is, the larger population migration will be.

Fourth, the new regional population for some small city i is obtained as

$$p_{1i} = p_i - p_{iE} + p_{iI}$$

Fifth, the new regional population path weight is recalculated by new path $\omega_i = l_i \Big/ \sum_{j=1}^{n} l_j$, where $l_j = p_{1i}/GDP_i$.

Five steps present the path-converged design of new regional population path after migration for small cities.

The migration here only concerns the interior migration among small cities. Step 4 needs to be updated for realization if the migration is implemented among small and large (medium) cities.

With the implementation of decision 1 for small cities, Fig. 3.13 and Table 3.6 illustrate the decision to promote urbanization for small cities in 2006. Figure 3.13 illustrates the rightward shift of the regional population path distribution with the decision to migrate. Table 3.6 shows that the inefficiency strength between the benchmark and regional population distributions shrinks to 0.05850 from 0.146 and that the average urbanization level experiences a growth of 1.840 percent (0.2370–0.2186). Both findings demonstrate that the decision to migrate from small cities is effective.

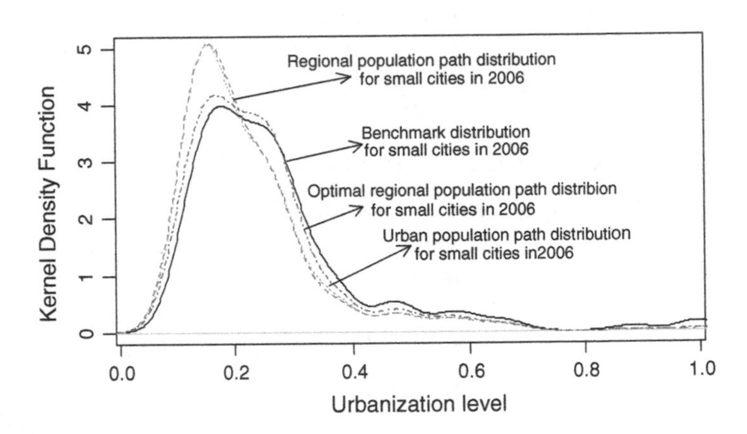

Fig. 3.13 New regional population path distribution for small cities in 2006

Table 3.6 Information regarding Strategy 1

Years	EX_1	Optimal EX_1	Inefficiency strength	Optimal inefficiency strength
2006	0.2190	0.2370	0.1460 (0.2140)	0.05850 (0.2780)

3.4.2 Strategy 2: Regional Migration for Medium Cities

l is 0.334 in the decision of regional population migration for medium cities.

First, the strength of migration population is 9.85 percent.

$$0.0985 = \int_{0.334}^{1} (f_n(x) - f_{\omega 1}(x))dx$$

Second, the total number of migration population is

$$0.0985 \sum_{k=1}^{n} p_k I(u_k < 0.334)$$

where p_k is the regional population of medium city k, and $\sum_{k=1}^{n} p_k I(u_k < 0.334)$ is the total regional population of medium cities whose urbanization level lower than 0.334.

Third, the emigration population for some medium city i is given by

$$p_{iE} = \frac{[f_{\omega_1}(u_i) - f_n(u_i)]I(u_i < 0.334)}{\sum_{j=1}^{n} [f_{\omega_1}(u_j) - f_n(u_j)]I(u_j < 0.334)} \times 0.0985 \sum_{k=1}^{n} p_k I(u_k < 0.334)$$

The immigration population for some medium city i is given by

$$p_{iI} = \frac{[f_n(u_i) - f_{\omega_1}(u_i)]I(u_i \geq 0.334)}{\sum_{j=1}^{n} [f_n(u_j) - f_{\omega_1}(u_j)]I(u_j \geq 0.334)} \times 0.0985 \sum_{k=1}^{n} p_k I(u_k < 0.334)$$

Fourth, the new regional population for some medium city i

$$p_{1i} = p_i - p_{iE} + p_{iL}$$

With reference to path-converged design, the implementation of strategy 2 for medium cities accompanies the information in Fig. 3.14 and Table 3.7. Figure 3.14 illustrates the rightward shift of the regional population path distribution with the decision to migrate. Table 3.7 illustrates the inefficiency strength between the benchmark and regional population distributions shrinks to 0.0412 from 0.0985 and the average urbanization level experiences a growth of 1.410 percent.

3.4.3 Strategy 3: Urban Migration for Medium Cities

l is 0.294 in the decision making of urban population migration for medium cities.

First, the strength of migration population is 7.10 percent.

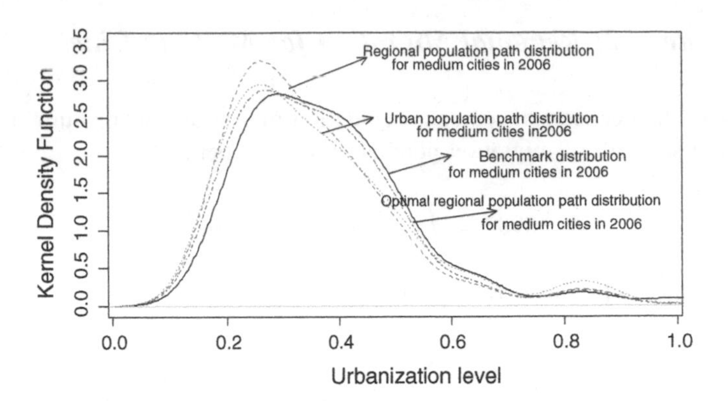

Fig. 3.14 New regional population path distribution for medium cities in 2006

Table 3.7 Information regarding Strategy 2

Years	EX_1	Optimal EX_1	Inefficiency strength	Optimal Inefficiency strength
2006	0.3408	0.3620	0.09850 (0.3340)	0.04120 (0.2950)

$$0.071 = \int_{0.294}^{1} (f_n(x) - f_{\omega_2}(x))dx$$

Second, the total number of migration population is

$$0.071 \sum_{k=1}^{n} p_k I(u_k < 0.294)$$

where p_k is the urban population of medium city k, $0.071 = \int_{0.294}^{1} (f_n(x) - f_{\omega_2}(x))dx$, and $\sum_{k=1}^{n} p_k I(u_k < 0.294)$ is the total urban population of medium cities whose urbanization level lower than 0.294.

Third, the urban emigration population for some medium city i is given by

$$p_{i\mathrm{E}} = \frac{[f_{\omega_2}(u_i) - f_n(u_i)]I(u_i < 0.294)}{\sum_{j=1}^{n} [f_{\omega_2}(u_j) - f_n(u_j)]I(u_j < 0.294)} \times 0.071 \sum_{k=1}^{n} p_k I(u_k < 0.294)$$

The immigration population for some medium city i is given by

$$p_{i\mathrm{E}} = \frac{[f_n(x_i) - f_{\omega_2}(x_i)]I(u_i \geq 0.294)}{\sum_{j=1}^{n} [f_n(x_j) - f_{\omega_2}(x_j)]I(u_j \geq 0.294)} \times 0.071 \sum_{k=1}^{n} p_k I(u_k < 0.294)$$

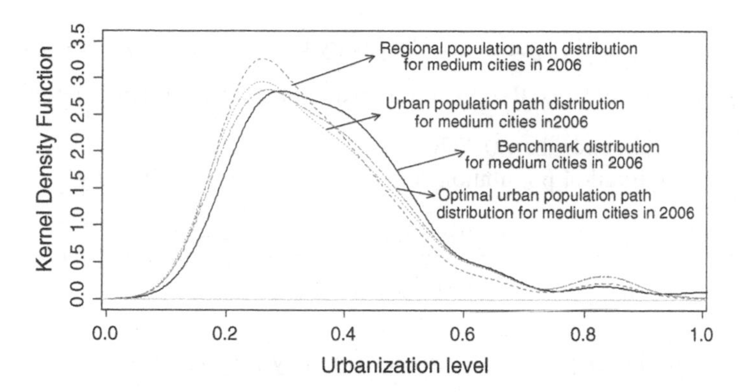

Fig. 3.15 New urban population path distribution for medium cities in 2006

Table 3.8 Information regarding strategy 3

Years	EX_2	Optimal EX_2	Inefficiency strength	Optimal Inefficiency strength
2006	0.364	0.3630	0.0710(0.294)	0.0473(0.2840)

Fourth, the new urban population for some medium city i

$$p_{1i} = p_i - p_{iE} + p_{il}$$

The implementation of decision No. 3 for medium cities presents the information in Fig. 3.15 and Table 3.8. Figure 3.15 illustrates the rightward shift of urban population path distribution with the decision making on population migration. Table 3.8 presents the inefficiency strength between the benchmark and urban population distributions shrinks to 0.0473 from 0.0710 and the average urbanization level experiences no obvious growth.

3.4.4 Strategy 4: Regional Migration for Large Cities

l is 0.492 in the decision of regional population migration for large cities.

First, the strength of migration population is 14.0 percent.

$$0.14 = \int_{0.492}^{1} (f_n(x) - f_{\omega_1}(x))dx$$

Second, the total of migration population is

$$0.14 \sum_{k=1}^{n} p_k I(u_k < 0.492)$$

where p_k is the regional population of large city k, $0.14 = \int_{0.492}^{1} (f_n(x) - f_{\omega_1}(x))dx$, and $\sum_{k=1}^{n} p_k I(u_k < 0.492)$ is the total regional population of large cities whose urbanization level lower than 0.4920.

Third, the emigration population for some large city i is given by

$$p_{iE} = \frac{[f_{\omega_1}(u_i) - f_n(u_i)]I(u_i < 0.492)}{\sum_{j=1}^{n} [f_{\omega_1}(u_j) - f_n(u_j)]I(u_j < 0.492)} \times 0.14 \sum_{k=1}^{n} p_k I(u_k < 0.492)$$

The immigration population for some large city i is given by

$$p_{iI} = \frac{[f_n(u_i) - f_{\omega_1}(u_i)]I(u_i \geq 0.492)}{\sum_{j=1}^{n} [f_n(u_j) - f_{\omega_1}(u_j)]I(u_j \geq 0.492)} \times 0.14 \sum_{k=1}^{n} p_k I(u_k < 0.492)$$

Fourth, the new regional population for some large city *i*

$$p_{1i} = p_i - p_{iE} + p_{iI}$$

The implementation of strategy 4 for large cities accompanies with the information of Fig. 3.16 and Table 3.9. Figure 3.16 experiences the rightward shift of the regional population path distribution with the decisions for population migration. Table 3.9 shows that the average urbanization level experiences a growth of 2.730 % and that the inefficiency strength between the benchmark and regional population distributions shrinks to 0.05620 from 0.1400.

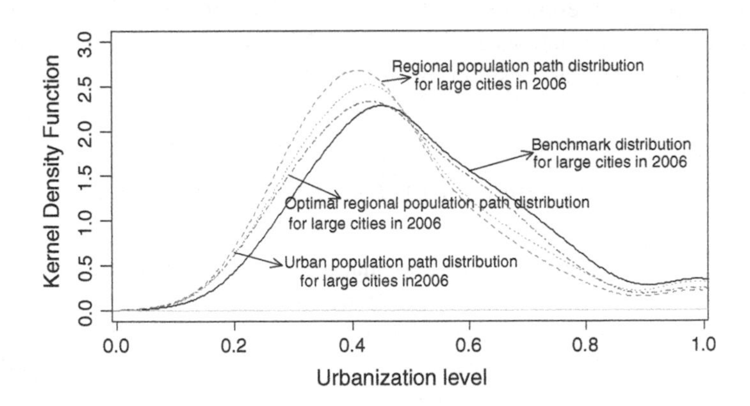

Fig. 3.16 New regional population path distribution for large cities in 2006

Table 3.9 Information regarding strategy 4

Years	EX_1	Optimal EX_1	Inefficiency strength	Optimal inefficiency strength
2006	0.4470	0.4740	0.1400(0.4920)	0.05620(0.4620)

3.5 Concluding Remarks

The study in this chapter continues the allocation efficiency framework, aims to identify and measure the allocation efficiency of population migration, taken as the labor input in growth system. Path-converged design identification is developed to identify the population agglomeration efficiency from both regional and urban perspectives, and to measure population agglomeration efficiency with proposed decisions of optimal population migration in urban planning in China.

3.5.1 Identification of Population Agglomeration Efficiency

First, the identification of urbanization level with benchmark model illustrates most cities locate at the urbanization level lower than 0.35, and the urbanization level grows gradually along with time because the *average growth rate* of urbanization level during 2002–2006 is 1.790 percent and the integral area at higher urbanization level *increases* gradually, in particular, the integral area at interval (0.7, 1) grows at 3.410 percent every year.

Second, the inefficiency of population agglomeration is identified for all cities in China. The identification of the population agglomeration shows that the population gathers asymmetrically around the average urbanization development level. Both regional and urban populations agglomerated to cities with urbanization levels lower than 0.35 during 2002–2006 because the integrals of both the regional population path and urban population path distributions at (0, 0.35) take up at least 75 percents for all years. Moreover, the agglomeration of both regional and urban population gradually shrinks over time because the integral areas at (0, 0.35) experiences a declining trend for both path distributions.

Third, the inefficiency of population agglomeration is also identified in large, medium and small cities. More specifically, the regional population gathers in cities with urbanization levels lower than 0.214 for small cities, lower than 0.3340 for medium cities, and lower than 0.4919 for large cities in 2006. Urban population agglomeration is very similar to regional population agglomeration in small and large cities; while the urban population in medium cities gathers in cities where urbanization is lower than 0.2940.

3.5.2 Decision Making of Optimal Population Migration

Based on the identification of population agglomeration inefficiency, four planning decisions regarding population migration are presented to eliminate inefficiency of population allocation.

Population migration is implemented based on the inefficiency strength of benchmark and population path distributions. The population migration number for any city is implemented according to the difference between the benchmark and original population path distributions; the greater the difference is, the larger population migration will be.

Strategy 1 for optimal regional population migration in small cities is to guide 14.64 percents of the regional population in small cities with urbanization levels lower than 0.214 to migrate to small cities at higher urbanization levels. Strategy 2 for regional population migration in medium cities is to guide 9.85 percents of the regional population in the medium cities with urbanization levels lower than 0.214 to migrate to medium cities at higher urbanization levels. Strategy 4 for regional population migration in large cities is to guide 14 percents of the regional population in the large cities with urbanization levels lower than 0.4919 to migrate to large cities at higher urbanization levels. The inefficiency strengths between benchmark and regional population distributions shrink to 5.800, 4.100, and 5.600 % from 14.64, 9.850, 13.97 % respectively, and average urbanization levels with regional population path distributions experience growths of 1.840, 1.410, and 2.730 percents for small, medium and large cities respectively. All findings show that the decisions regarding regional population migration are effective.

However, the population migration this study employed is the interior migration for small, medium and large cities respectively. The path-converged design needs to be updated for realization on migration if the population migration is implemented among different city groups, which will definitely bring about difference in conclusions.

References

Cai, J. (2006). Institutional analysis for the lag of urbanization behind the economic development. *City Planning Review, 1*, 67–72.

Chen, Q. (2006). Crisis and opportunities of the urbanization in China. *City Planning Review, 1*, 34–39.

Henderson, J. V. (2003). The urbanization process and economic growth: the so-what question. *Journal of Economic Growth, 8*, 47–71.

Johnson, D. G. (2004). Provincial migration in China in the 1990s. *China Economic Review, 14*, 22–31.

Li, L. J., & Jin, J. (2005). Urbanization level forecast time series model and application in China. *China's Population Science*, (*12*), 1–6.

Lu, M., & Chen, Z. (2004). The economic policy of urbanized, Urban tendency and income difference between city and countryside. *Economic Research Journal, 6*, 41–49.

Maria, G. P., & Roberto, Z. (2004). Testing for changing shapes of income distribution: Italian evidence in the 1990s from kernel density estimates. *Empirical Economics, 29*, 415–430.

Silverman, B. W. (1981). Using kernel density estimates to investigate multi-modality. *Journal of Royal Statistical Society (B), 43*, 97–99.

Wang, G. X., & Huang, Y. Y. (2005). Population migration and eastern region economic development of chinese interprovincial, 1995–2000. *Population Research, 1*, 19–28.

Xu, B., & Watada, J. (2006). Dynamic of the Chinese urbanization base on nonparametric models. In: Proceeding of the 7th International Conference on Intelligent Technologies, Vol. 14, pp. 162–173.

Xu, B., & Watada, J. (2007). 2007a. *Nonparametric Identification of GDP with urbanization: A Study in China, ICIC express letters, 1*(1), 73–80.

Xu, B., & Watada, J. (2008a). Identification of regional urbanization gap: Evidence of China. *Journal of Modeling in Management, 3*(1), 7–25.

Xu, B., & Watada, J. (2008b). Observed probability measurement for urbanization development level with errors-in-variables observation. *International Journal of Innovative Computing, Information and Control, 4*(5), 1233–1242.

Xu, B., Zeng, J., & Watada, J. (2008). Distribution identification and path selection for population agglomeration in Chinese urbanization development. *International Journal of Intelligence Technologies and Applied Statistics, 1*(2), 55–74.

Yang, Y. (2003). Scale reckoning and strength analysis of chinese population migration. *Social Science in China, 6*, 97–107.

Yu, L., & Zhang, S. Y. (2006). The migration flow and the affecting variables: A case study of the three megalopolises of China. *South China Population, 3*, 17–23.

Chapter 4
Changes in Factors Scale Efficiency of FDI with Path Identification

Besides the direct change of production efficiency in the channel of allocation efficiency in Chaps. 2 and 3, the FDI absorption and domestic private investment will inevitably bring about indirect impact effect on production efficiency. The first part of indirect change in production efficiency refers to the change in scale efficiency, which reflects whether the additional investment will bring about crowd-in or crowd-out effect on initial endogenous capital and furthermore scale expansion or scale downsizing on the production efficiency of both capital and labor ($\alpha + \beta$) in Eq. (1.7).

4.1 Crowd-in or Crowd-out Effects of FDI Scale

Two contrary decision makings are presented on whether FDI should be encouraged to reduce regional discrepancy between Eastern and Western China by Wei (2002) and Wu (2002) respectively. In this study, the *Path-identification* approach is established and applied to review the question with the aim to propose reasonable decision makings.

FDI contributed to economic growth of the host economy through learning, diffusion of technology, positive externalities and capital inflows (Lai et al. 2006; Liu 2008; Li and Liu 2005; Luo 2007). How FDI affects regional growth disparity has also been the focus of public attention (Ma and Guo 2007; Ng and Tuan 2006; Ma 2006; Ran et al. 2007), the governmental decision makers are confused on whether FDI should be encouraged in Middle and Western regions with the aim of reducing regional discrepancy since previous researches present divergences on to what extent FDI is responsible for the regional discrepancy in China.

More Specifically, Wei holds that about 90 % of China's regional discrepancy in growth can be explained by the unbalanced regional distribution of FDI in Chinese authoritative *Economic Research Journal,* with the following model employed:

$$Y = AL^{\beta}K_G^{\alpha}K_P^{\rho}K_F^{\lambda} \tag{4.1}$$

B. Xu et al., *Changes in Production Efficiency in China*,
DOI: 10.1007/978-1-4614-7720-4_4,

where Y is output, L is the labor, K_G is the domestic governmental investment, K_P is the domestic private investment, K_F is the foreign direct investment, and A refers to the productive technology.

In the same *Economic Research Journal,* Wu (2002) points out that less than 20 % of regional discrepancy is attributed to the regional differences of FDI, with the following model employed:

$$Y = AL^{\beta} K_D^{\alpha} K_N^{\rho} K_F^{\lambda} \tag{4.2}$$

where K_D is the domestic investment, K_N is the non-investment factor (e.g. natural resource).

Accordingly, Wei (2002) puts forward that energetic efforts should be made to actively absorb foreign capital into Middle and Western regions so as to push their developments; while Wu (2002) denies the possibility that the FDI regional distribution change could narrow the regional discrepancy in China.

The completely different conclusions and proposals by Wei (2002) and Wu (2002) confuse the comprehension on whether regional imbalance of FDI will result in regional growth discrepancy. Qian (2007) compares the two studies and holds that the secondary error could be formed during experimental researches in aspects of variables selection, model establishment and data application. Unfortunately, Qian (2007) fails in explaining the key to the rise of the question and comes to an unacceptable conclusion that there is at least one incorrect conclusion between them, which challenges the studies by Wei (2002) and Wu (2002).

Similar to the researches of Wei (2002) and Wu (2002), taken as *Factor-Driven Growth approach*, the generalized C-D production function is obtained by adding related factors into the C-D function. Fedderke and Romm (2006) include the stock of FDI, Goss et al. (2007) refer to the labor quality, Li and Liu (2005) include the secondary school attainment level, Ramirez (2006) adds government consumption expenditures, the real exports and other political events rather than the human capital, Hu (2007) introduces urbanization, R&D level, technological resources into the CD production function to seek the relationship between FDI and economic growth. Besides, different proxy variables are used to substitute the same input factor. The addition of different related factors to C-D production function and substitution of the same factor with different proxy variables will inevitably bring about diverse conclusions on the same question.

Actually, Fine (2000) mentions the following serious problems the econometric models encounters with. First, the independent variables in the context will inevitably be related to another one, since the correlations of growth are systematically connected, quite apart from the mutuality of dependent and independent variables. For example, foreign capital has crowd-in or crowd-out effect on native capital in China (Xin and Deng 2007), so the variable K_F can't be independent of variables K_D, K_N, K_P in models (4.1) and (4.2). Second, the econometrics is highly selective in terms of the relations that it does examine as opposed to those that it does not. For example, both models (4.1) and (4.2) concern with output growth, which model should be selected to explore how regional discrepancy attributes to FDI since the respective λ and variables are different in the models (4.1) and (4.2)? Third, it is important to stress how stochastic variation is generated in the model since, according to endogenous growth theory, random shocks can have a persistent and varied impact over time, depending upon how they arise and how they are transmitted. i.e., λ should vary along with time rather than keep constant in models (4.1) and (4.2).

Compared with previous *Factor-Driven growth approach*, this research puts forward an innovative *Path identification approach* in terms of *Decision-Making* firstly to identify production discrepancy due to FDI among regions in China, secondly to unity the two seemingly contrary conclusions that regional discrepancy of FDI explains respectively about less than 20 % (Wu 2002) and almost 90 % (Wei 2002) of development discrepancy between Eastern and Western regions in China, and finally to develop environment FDI path identification to design reasonable decision makings for reducing region discrepancy among regions.

The further study is organized as follows. Section 4.2 provides the nonparametric identification technique Sect. 4.3 puts forward the specific FDI Path *identification* approach. Section 4.4 presents the data and empirical results of scale efficiency with FDI Path *identification.* Section 4.5 measures the investment strategy with urbanization environment to realize the crowd-in effect. Section 4.6 concludes.

4.2 Nonparametric Identification Technique

Nowadays, the study on identification has been broadly associated with the design of experiments. In biological and physical sciences, an investigator who wishes to make inferences about certain parameters can usually conduct controlled experiments to isolate relations. However, the cases in management engineering on social sciences are less fortunate. Certain facts can be observed and it is naturally required to arrange them in a meaningful way for studies. Yet their natural conditions usually can not be reproduced in laboratory. It is hard to control the variables and isolate the relations and it is almost impossible to identify an unknown latent system that produces data used in the analysis. However, fortunately in the view of management engineering on social sciences, the identification focuses on the change effects of the underlying structure generated by controlled conditions rather than the underlying structure itself. The study in this chapter aims to build an identification framework of change effects due to the underlying structure.

Let x, y be a set of such observations. Structure S denotes the complete specification of probability distribution function, $F(x, y|S)$ of x and y. The set of all ***structure*** S is called a ***system*** $\mathbb{S}$. A possible prior structures S_0 of the ***system*** $\mathbb{S}$ is called a ***model***. The identification problem is to decide the judgment about ***model***, given structure S and observations x and y.

In the case of parametric identification, it is assumed that x and y are generated by parametric probability distribution function, that is

$$F(x, y|S_0) = F(x, y|\alpha) \tag{4.3}$$

where α is an m-dimensional real vector. Probability distribution function F is assumed known, which is conditional on α. However, parameter α is unknown. Hence, a structure is described by a parametric point α, and a model is described by a set of points $A \subset R^m$. Thus, the problem of distinguishing between structures is reduced to the problem of distinguishing between parameter points. In essence, this parameter identification can be solved through the estimation of unknown parameter α using statistical method.

In the case of nonparametric estimation, probability distribution function F is unknown. Given actual data x and y, the nonparametric estimations can be obtained for example, conditional density $f(y|x)$, regression function $m(x) = E(Y|x)$, etc. The rationality of estimations is proved by the convergence of the nonparametric estimated $\hat{F}$ to the unknown F. In the case of nonparametric test, probability distribution function F is assumed to equal to a known function F_0, the nonparametric test is implemented by testing the null hypotheses $F = F_0$ versus $F \neq F_0$.

There has been nonparametric identification that only handles some dichotomous classification, such as the identification of regression models with misclassification on dichotomous regressor (Mahajan 2006; Chen et al. 2007), the nonparametric identification of the classical errors-in-variables model (Schennach et al. 2007) and semiparametric identification of structural dynamic optimal stopping time models (Chen 2007). Based on the ideas of parameter identification and nonparametric estimation, the book brings forward original nonparametric ***Path*** identification. Different from dichotomous classification, the identification strategy in this book focuses on the nonparametric path identification of an underlying structure.

Assume real-world data x, y are generated by nonparametric probability distribution function $F(x, y|S_0)$, where the probability distribution function F is unknown, and S_0 is called as an ***underlying structure***. No parameters in F need to be estimated or dichotomous classification needs to be identified. In this case, it is impossible to reduce the problem of distinguishing between structures to the problem of distinguishing between parameter points. The concern is how to identify underlying structure S_0. The original identification of underlying structure S_0 with probability one can be achieved by the following three steps.

Step 1. ***Benchmark Model***: **Estimating underlying structure**

Using the nonparametric approach, the unknown probability distribution function$F(x, y|S_0)$ can be estimated based on observed data x and y. For example, conditional density $f(y|x)$, regression function $m(x) = E(Y|x)$, etc.

Let $F_n(x, y|S_1)$ is an estimation of $F(x, y|S_0)$, where n is the sample size of (x, y), S_1 is called as a ***benchmark model***.

Intuitively, taken the view of statistics, the following assumption needs to be satisfied. The nonparametric estimation, $F_n(x,y|S_1)$, is the strong consistent estimation of $F(x,y|S_0)$, i.e.

$$P\left(\lim_{n\to\infty} F_n(x,y|S_1) = F(x,y|S_0)\right) = 1 \tag{4.4}$$

where P is a probability measurement.

Step 2. ***Path Model :*** **Finding controllable factors**

Imitating the identification of parameter α in parameter identification, this study introduces an new observation variable Z into the probability distribution function$F(x,y)$ to obtain $F_n(x,y|Z)$. Recalling the law of diminishing marginal returns to Z *in view of economics*, and the law of large numbers of Z *in view of statistics*, it gives that restriction of a formation Z is indispensable. The variable Z is taken as a path model, if the following equation is satisfied:

$$P\left(\lim_{n\to\infty}[F_n(x,y|Z) - F_n(x,y|S_1)] = 0\right) = 1 \tag{4.5}$$

This equation in the book can be call as Z ***Path-Converged Design***. It is of crucial importance to the original nonparametric identification.

Compared with parameter identification of (4.3), it has ***Path model*** S_2 identifiable by path Z, i.e., there is $F_n(x,y|S_2)$, such that

$$F_n(x,y|S_2) = F_n(x,y|Z) \tag{4.6}$$

In other words, path structure S_2 can be found uniquely through observation variable Z.

According to parameter identification of (4.3), the parameter α is unique. However, observation Z is unique only with probability one in (4.6). Hence, Z is not a unique identification of ***underlying structure*** S_0.

Compared with constant parameter α, observation Z may be a comprehensive integration set of various factors, which reflects the dynamic change of ***underlying structure*** S_0. Although the known observation Z is not a full identification of underlying structure S_0 itself, the Z identification focuses on the change effects generated by the underlying structure rather than the underlying structure itself. However, the change effects of S_0 identified by Z is not given explicitly since S_0 is unknown.

Step 3. ***Path Identification***

Now, the problem of identification of underlying structure S_0 is reduced to the problem of identification of the change effects generated by the ***Path model*** S_2 and ***Benchmark model*** S_1.

Through the path Z, the identification of underlying structure $F(x,y|S_0)$ is reduced to the identification of change effects between benchmark model. $F_n(x,y|S_1)$ and path model $F_n(x,y|S_2)$.

In fact, combine Eq. (4.4) with Eq. (4.5) and get

$$P\left(\lim_{n\to\infty} F_n(x,y|Z) = F(x,y|S_0)\right) = 1$$

Hence, according to parameter identification of (4.3), underlying structure S_0 is identifiable by the observed path Z, with probability one.

The following definition states the nonparametric Z path identification with probability one.

Definition 4.1 Underlying structures S_0 is said to be *Z* ***path identifiable*** with probability one, if the following equations are satisfied:

1. $P\left(\lim_{n\to\infty} F_n(x,y|S_1) = F(x,y|S_0)\right) = 1$
2. $P\left(\lim_{n\to\infty} [F_n(x,y|S_2) - F_n(x,y|S_1)] = 0\right) = 1$

where $F_n(x,y|S_2) = F_n(x,y|Z)$. The identification means that path Z realizes the identification of underlying structure S_0 by observing the change effects of path model S_2 based on benchmark model S_1 with probability one.

Figure 4.1 illustrates the main idea of nonparametric path identification.

Two remarks are given considering the path identification framework in this book.

Remark In Chap. 2, the ω ***path identification*** is established based on kernel density function of production structure. Using the theoretical nonparametric ***Path*** identification, it can be concluded that ω ***path identification*** **is a** specific example of nonparametric Path identification only when the probability distribute function *F*in Definition 4.1 is replaced by the probability density function *f*. The key of path identification in Path-Converged Design is that the observed variables depend not only on sample number, but also on the time.

Remark Fine (2000) mentions the econometrics of endogenous growth has confronted with three problems, that is, the problems of time-varying feature, model selection and independence assumption. The time-varying elasticity production function introduced in this chapter (Xu and Watada 2007b) has settles the first problem. Given a benchmark structure of driven growth, the nonparametric

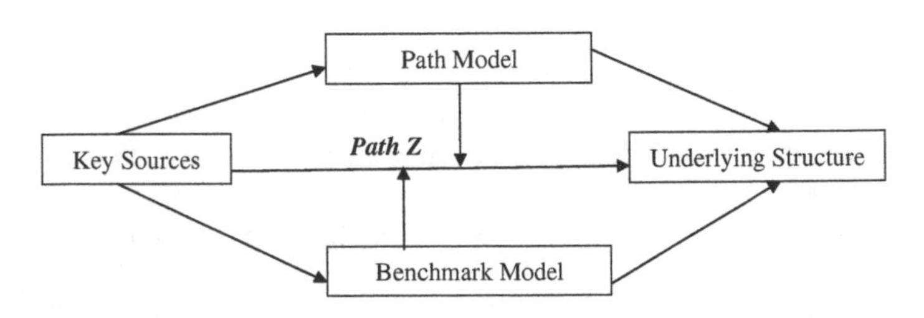

Fig. 4.1 Nonparametric ***Path*** identification

identification of underlying structure is obtained by comparison of benchmark with path structures. Therefore, the identification handles the model selection issue by comparison with a given benchmark structure. Moreover, the comparison successfully avoids the issue of independence assumption.

4.3 FDI Path Identification

The diverse conclusions are probably permissive and significant in terms of both Economics and Statistics. Taken the studies of Wei (2002) and Wu (2002) for instance, on the one hand, in terms of economics, model (4.1) divides the domestic investment into government investment and private investment; model (4.2) employs a single domestic investment factor, and a non-investment productive (natural source) factor for analysis. Both researches by models (4.1) and (4.2) are meaningful and feasible. On the other hand, in terms of statistics, both models are statistically significant since respective statistical hypothesis test of model is accepted in each study.

However, the diverse conclusions of 90 and 20 % puzzle whether the FDI inflow in Eastern region should be encouraged with the aim of eliminating regional discrepancy. Since the question can't be settled perfectly simultaneously with the perspectives of Economics and Statistics, the study aims to employ the idea of Decision Making to settle the confused and important question.

The study in this section puts forward FDI path identification. It is a specific example of nonparametric ***Path*** identification when the degenerative probability distribution function F is presented in ***Definition 4.1***. It is established by the following three steps.

Step 1. ***Benchmark model***

Assume the well-known C-D production function as equilibrium of economic system, which is presented by the equilibrium structure S_0:

$$Y(t) - A_0(t)K(t)^{\alpha}L(t)^{\beta} = 0 \tag{4.7}$$

where K and L denotes the inputs of labor and capital, α and β are given in (4.21) below and correspond to output elasticity of capital and elasticity of labor respectively, A_0 refers to the technical level of production.

Since the initial equilibrium structure S_0, taken as the underlying structure, is unknown, the first step for further research is to estimate S_0, which is obtained as follows.

Define a benchmark model S_2 by:

$$Y_2(t) - K(t)^{\alpha(t)}L(t)^{\beta(t)} = 0 \tag{4.8}$$

Intuitively, taken the view of statistics, the following assumption needs to be satisfied for study:

$$P\left(\lim_{t\to\infty}[Y_2(t) - Y(t)] = 0\right) = 1 \tag{4.9}$$

Assumption (4.9) indicates that $Y_2(t)$ is a strong consistent estimator of $Y(t)$.

$$Y_2(t) - K(t)^{\alpha(t)}L(t)^{\beta(t)} = 0 \tag{4.10}$$

It is a time varying production function, $\alpha(t)$ and $\beta(t)$ are time-varying elasticity of capital and labor at time t respectively, which can be estimated by the non-parametric local polynomial regression method and derivative estimation.

Step 2. ***Path model*** S_2

Now, suppose a new exterior factor FDI enters into the initial equilibrium structure S_0, it is taken as FDI path in the study. The change process from equilibrium production structure S_0 to the disequilibrium production model S_1 is represented as follows:

$$Y(t) - FDI(t)^{\lambda(t)}K(t)^{\alpha(t)}L(t)^{\beta(t)} \neq 0 \quad if \quad FDI \neq\equiv 1 \tag{4.11}$$

Since FDI is a new factor that introduced in the initial equilibrium, the change process from equilibrium to disequilibrium structure also presents reallocation efficiency of the new factor FDI.

To depict the change process, define the path model S_1 by:

$$Y_1(t) - FDI(t)^{\lambda(t)}K(t)^{\alpha(t)}L(t)^{\beta(t)} = 0 \tag{4.12}$$

where FDI refers to the Path FDI, $\lambda(t)$ is the time-varying elasticity of Path FDI.

Suppose $FDI(t) = 1$ which return to (4.8) denotes static equilibrium, then $FDI(t)$ in (4.11) is an indicator of the deviation degree of dynamic disequilibrium from static equilibrium in (4.8). Coefficient $\lambda(t)$ represents the strength of disequilibrium.

Suppose that $FDI(t) = 1$ defines a dynamic equilibrium relationship among all t, then FDI (t) is an indicator of the degree of disequilibrium. Coefficient $\lambda(t)$ represents the strength of disequilibrium and the Path model is now said to be in *Path-converged* form by FDI.

Step 3. ***Path convergence design***

Now, the problem is how to specify the function of $FDI(t)$ to identify the $FDI(t)$ efficiency. Theoretically, $FDI(t)$ is a stochastic process. A sample path for $FDI(t)$ is the function on t to the range of the process which assigns to each t the value $FDI(t)$, where t is a previously given fixed point in the domain of the process.

Recalling the law of diminishing marginal returns to $FDI(t)$ *in view of economics*, it gives

$$P\left\{\lim_{t\to\infty}\frac{\partial FDI(t)}{\partial t} = 0\right\} = 1 \tag{4.13}$$

And recalling the law of large numbers of $FDI(t)$ *in view of statistics*, it gives

$$P\left\{\lim_{n\to\infty}\frac{1}{n}\sum_{t=1}^{n}(FDI(t)-EFDI(t))=0\right\}=1 \tag{4.14}$$

Equations (4.13) and (4.14) illustrate the path behavior of $FDI(t)$ which will converge to constant when $t\to\infty$, or $n\to\infty$.

The theories in both economics and statistics do not present specified function of $FDI(t)$. However, combining (4.9) with (4.13), (4.14) above, they bring about the analysis in view of engineering as follows.

The final process, after taking some time to adjust the disequilibrium (4.11) into re-equilibrium of resource allocation (4.8), involved is how to model the changing process of the production structure *from the disequilibrium to equilibrium.*

The specified function of $FDI(t)$ will be designed as follows with the help of re-equilibrium production structure.

This book will design a new sample path $fdi(t)$, a function of $FDI(t)$, to replace $FDI(t)$, such that re-equilibrium production structure can be achieved as follows

$$P\left(\lim_{t\to\infty}[Y_1(t)-Y_2(t)]=0\right)=1 \tag{4.15}$$

In the change process from equilibrium to disequilibrium, and disequilibrium to equilibrium production structure, it achieves identification of the change effects of underlying production structure generated by the path model converge to benchmark model along with $FDI(t)$ path.

In terms of convergence behavior, the path $fdi(t)$ is called as ***Path-converged design***.

The applications of (4.9) and (4.15), the efficiency identification of $FDI(t)$ is obtained by:

$$Y_1(t)-Y_2(t) \tag{4.16}$$

Definition 4.2 Underlying structures S_0 is said to be FDI ***path identifiable*** through Path model S_2 with probability one, if the following conditions are satisfied,

$$P\left(\lim_{t\to\infty}[Y_1(t)-Y_0(t)]=0\right)=1 \tag{4.17}$$

$$P\left(\lim_{t\to\infty}[Y_2(t)-Y_1(t)]=0\right)=1 \tag{4.18}$$

The identification here referring to Path model S_2 realizes the identification of underlying structure S_0 with probability one by observing the change effects of Path model S_2 based on Benchmark model S_1.

Now, functions $\lambda(t)$, $\alpha(t)$ and $\beta(t)$ are assumed the same degrees of smoothness and hence they can be approximated equally well at the same interval. Use local linear modeling to estimate coefficient function (4.10) and (4.12) respectively. Suppose t locates at a neighborhood of t_0, the local approximations are given as:

$$\lambda(t) = \lambda(t_0) + \lambda'(t_0)(\mathrm{t} - \mathrm{t}_0) + \mathrm{o}(\mathrm{t} - \mathrm{t}_0) = \lambda_0 + \lambda_1(\mathrm{t} - \mathrm{t}_0) + \mathrm{o}(\mathrm{t} - \mathrm{t}_0)$$

$$\alpha(t) = \alpha(t_0) + \alpha'(t_0)(\mathrm{t} - \mathrm{t}_0) + \mathrm{o}(\mathrm{t} - \mathrm{t}_0) = \alpha_0 + \alpha_1(\mathrm{t} - \mathrm{t}_0) + \mathrm{o}(\mathrm{t} - \mathrm{t}_0)$$

$$\beta(t) = \beta(t_0) + \beta'(t_0)(\mathrm{t} - \mathrm{t}_0) + \mathrm{o}(\mathrm{t} - \mathrm{t}_0) = \beta_0 + \beta_1(\mathrm{t} - \mathrm{t}_0) + \mathrm{o}(\mathrm{t} - \mathrm{t}_0)$$

where $x'(t) = dx/dt$ is the time derivatives of respective variable, and $\lim_{t \to t_0} o(\mathrm{t} - \mathrm{t}_0) = 0$.

Thus function (4.10) and (4.12) can be expressed respectively as

$$\begin{aligned} Y_1(t) &= K(t)^{\alpha(t)} L(t)^{\beta(t)} \\ &= K(t)^{\alpha_1(t)} L(t)^{\beta_1(t)} K(t)^{\alpha_0 - \alpha_1 t_0} L(t)^{\beta_0 - \beta_1 t_0} (K(t)L(t))^{o(t - \mathrm{t}_0)} \\ &= A_1(t) K(t)^{\alpha} L(t)^{\beta} (K(t)L(t))^{o(t - \mathrm{t}_0)} \end{aligned} \tag{4.19}$$

and

$$\begin{aligned} Y_2(t) &= FDI(t)^{\lambda(t)} K(t)^{\alpha(t)} L(t)^{\beta(t)} \\ &= FDI(t)^{\lambda_1(t)} K(t)^{\alpha_1(t)} L(t)^{\beta_1(t)} FDI(t)^{\lambda_0 - \lambda_1 t_0} K(t)^{\alpha_0 - \alpha_1 t_0} L(t)^{\beta_0 - \beta_1 t_0} (FDI(t)K(t)L(t))^{o(t - \mathrm{t}_0)} \\ &= A_2(t) FDI(t)^{\lambda} K(t)^{\alpha} L(t)^{\beta} (FDI(t)K(t)L(t))^{o(t - \mathrm{t}_0)} \end{aligned} \tag{4.20}$$

where

$$\alpha = \alpha_0 - \alpha_1 t_0,\ \beta = \beta_0 - \beta_1 t_0,\ \lambda = \lambda_0 - \lambda_1 t_0 \tag{4.21}$$

$$A_1(t) = K(t)^{\alpha_1(t)} L(t)^{\beta_1(t)},\ A_2(t) = FDI(t)^{\lambda_1(t)} K(t)^{\alpha_1(t)} L(t)^{\beta_1(t)} \tag{4.22}$$

are technical level of Benchmark model and Path model respectively.

To validate the underlying structure S_0 is identifiable; the following proves both assumptions (4.17) and (4.18) can be satisfied.

Suppose the following probability relations hold true:

$$P\left(0 < \lim_{t \to \infty} K(t)L(t) < \infty\right) = 1, \quad P\left(0 < \lim_{t \to \infty} A_0(t) < \infty\right) = 1 \tag{4.23}$$

$$P\left(0 < \lim_{t \to \infty} [A_1(t) - A_0(t)] = 0\right) = 1 \tag{4.24}$$

First, assumption (4.23) is true, because production technical level, the capital and labor inputs always keep positive and finitive.

Secondly, in view of economics, FDI_t and GDP_t both increase with time t in probability one. Replace FDI_t by

$$fdi_i = \frac{(n-i+1)b_i}{\sum_{j=1}^{n} b_j}$$

where $b_1 = FDI_1/GDP_1$ $b_t = FDI_t/GDP_k$, $t = 2$, $k = \max\{i : 1 \geq FDI_t/GDP_i \geq b_{t-1}, i = 1, \ldots, t\}$.

It has b_t increasing with time t, and

$$P\left(\lim_{t\to\infty} b_t = constant\right) = 1$$

Furthermore

$$P\left(\lim_{t\to\infty} fdi_t = 1\right) = 1 \tag{4.25}$$

Thirdly, Eq. (4.24) is explicit since large sample theory of statistics two technical levels difference will gradually diminish and eventually disappear with probability one as $t \to \infty$.

Theorem 4.1 *Underlying structure S_0 can be identified under Assumptions (4.23) and (4.24).*

Proof Using (4.19) and (4.20), it follows that

$$P\left(\lim_{t\to\infty}\left[K(t)^{\alpha(t)}L(t)^{\beta(t)} - A_0(t)K(t)^{\alpha}L(t)^{\beta}\right] = 0\right)$$

$$P\left(\lim_{t\to\infty}\left[A_1(t)K(t)^{\alpha}L(t)^{\beta}(K(t)L(t))^{o(t-t_0)} - A_0(t)K(t)^{\alpha}L(t)^{\beta}\right] = 0\right)$$

$$P\left(\lim_{t\to\infty}\left[A_1(t)K(t)^{\alpha}L(t)^{\beta}(K(t)L(t))^{o(t-t_0)} - A_0(t)K(t)^{\alpha}L(t)^{\beta}\right] = 0, 0 < \lim_{t\to\infty} K(t)L(t) < \infty\right)$$

$$P\left(\lim_{t\to\infty}\left[A_1(t)(K(t)L(t))^{o(t-t_0)} - A_0(t)\right] = 0\right)$$

$$P\left(\lim_{t\to\infty}\left[A_1(t)(K(t)L(t))^{o(t-t_0)} - A_0(t)\right] = 0, \lim_{t\to t_0}(K(t)L(t))^{o(t-t_0)} = 1\right)$$

$$+P\left(\lim_{t\to\infty}\left[A_1(t)(K(t)L(t))^{o(t-t_0)} - A_0(t)\right] = 0, \lim_{t\to t_0}(K(t)L(t))^{o(t-t_0)} \neq 1, 0 < \lim_{t\to\infty} K(t)L(t) < \infty\right)$$

$$+P\left(\lim_{t\to\infty}\left((K(t)L(t))^{o(t-t_0)}\right)[A_1(t) - A_0(t)] + \lim_{t\to\infty} A_0(t)\left[(K(t)L(t))^{o(t-t_0)} - 1\right] = 0, \lim_{t\to t_0}\left((K(t)L(t))^{o(t-t_0)}\right) = 1\right)$$

$$= P\Big(\lim_{t\to\infty}[A_1(t) - A_0(t)] + \lim_{t\to\infty} A_0(t)\Big[(K(t)L(t))^{o(t-t_0)} - 1\Big]$$
$$= 0, \lim_{t\to t_0}\Big((K(t)L(t))^{o(t-t_0)}\Big) = 1, 0 < \lim_{t\to\infty} A_0(t) < \infty\Big)$$

$$= P\Big(\lim_{t\to\infty}[A_1(t) - A_0(t)] = 0\Big) = 1 \tag{4.26}$$

and

$$P\Big(\lim_{t\to\infty}\Big[FDI(t)^{\lambda(t)}K(t)^{\alpha(t)}L(t)^{\beta(t)} - K(t)^{\alpha(t)}L(t)^{\beta(t)}\Big] = 0\Big)$$

$$= P\Big(\lim_{t\to\infty}\Big[FDI(t)^{\lambda(t)} - 1\Big] = 0\Big) = 1$$

$$= P\Big(\lim_{t\to\infty}\Big[FDI(t)^{\lambda_0+\lambda_1(t-t_0)+o(t-t_0)} - 1\Big] = 0\Big) = 1$$

$$= P\Big(\lim_{t\to\infty}\Big[FDI(t)^{\lambda_0+\lambda_1(t-t_0)+o(t-t_0)} - 1\Big] = 0, \lim_{t\to\infty} FDI(t)$$
$$= 1, \lim_{t\to t_0}|\lambda_0 + \lambda_1(t-t_0) + o(t-t_0)| < \infty\Big) = 1 \tag{4.27}$$

□

According to ***Definition 4.2***, underlying structure S_0 can be path-identified with Path model S_2 based on Benchmark model S_1.

Models (4.12) can be estimated by the local weighed least-squares technique:

$$\min_{\theta(t_0)}\sum_{t=1}^{n}(\ln Y_t - (\lambda_0 + \lambda_1(t-t_0))FDI_t - (\alpha_0 + \alpha_1(t-t_0))\ln K_t - (\beta_0 + \beta_1(t-t_0))\ln L_t)^2 K_h(t-t_0) \tag{4.28}$$

where $\theta(t_0) = (\lambda_0, \alpha_0, \beta_0, \lambda_1, \alpha_1, \beta_1)^T$, $K_h(x) = h^{-1}K(x/h)$ and h are kernel functions and bandwidth.

The bandwidth h in this research is selected by the least squares cross-validation method (Stone 1984), and kernel function is given by Gauss kernel function.

4.4 Data Description and Empirical Identification of FDI

4.4.1 Data Description

The data source is China Compendium of Statistics 1949–2004, China Statistical Yearbook 2006, Statistical Communique on 2005 National Economic and Social Development of each Province in China. The China Compendium of Statistics

1949–2004 presents panel data of regional Gross Domestic Product (GDP), Indices of Gross Domestic Product (1952 = 100), number of year-end employees (Labor), amount of foreign direct investment actually used (FDI) during 1985–2004 for 31 provinces (including autonomous regions, and municipalities directly under the Central Government). China Statistical Yearbook 2006 provides the GDP, GDP indices and labor data for each province in 2004. The Statistical Communique on 2005 National Economic and Social Development of each Province gives the amount of foreign direct investment actually used (FDI) in each province. Zhang (2008) presents the panel data of real capital stock during 1985–2005 for 30 provinces at the 1952s price, which is calculated by the perpetual inventory approach.

Since some provinces lack the data of FDI or capital stock, only 29 provinces are included in the research, with Haina omitted in Eastern region and Chongqing and Xizang omitted in Western region. The variables GDP, real capital stock, labor and FDI path have a long-term conintegration relationship during 1985–2004 in each region.

4.4.2 Identification of Crowd-in and Crow-out Effects of FDI Path

Given Models (4.10) and (4.12) estimated by local weighed least-squares tech nique, the identification of whether FDI path contributes the regional discrepancy through both perspectives of capital elasticity and output efficiency.

4.4.2.1 Capital Elasticity: FDI Path Crowd-in or Crowd-out

The capital elasticity curves are listed in Figs. 4.2, 4.3, 4.4. "$K1$", "$K2$" and "$K3$" are the capital elasticity of FDI path model in Eastern, Middle and Western regions respectively, and "$K01$", "$K02$" and "$K03$" are the time-varying capital elasticity of benchmark model in each region.

Definition 4.3 FDI Path has crowd-in effect for the capital factor at time t if underlying structure S_0 is FDI ***path identifiable*** and $K1(t) \geq K01(t)$, otherwise, $K1(t) \geq K01(t)$, FDI Path has crowd-out effect for the capital factor.

Here crowd-out or crowd-in effect refers to the decrease or increase effect of the utilization efficiency of capital comparing FDI path model with benchmark model.

1. **Capital elasticity**.

Capital elasticity witnesses crowd-out (decrease) effect during 1989–1995 and crowd-in (increase) effect after 1996 with FDI in Eastern region. Unfortunately, the capital elasticity witness the crowd-out (decrease) effect in both Middle and

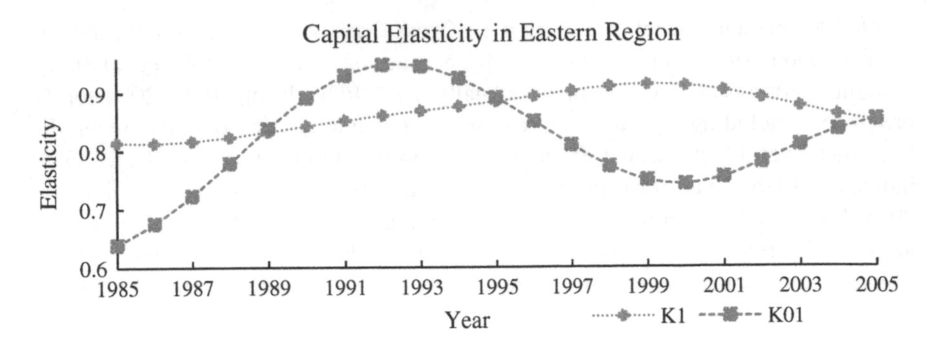

Fig. 4.2 The capital elasticity in eastern region

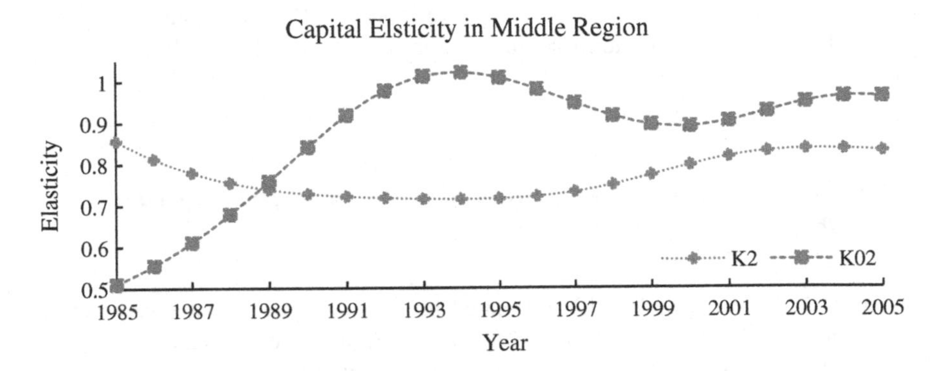

Fig. 4.3 The capital elasticity in middle region

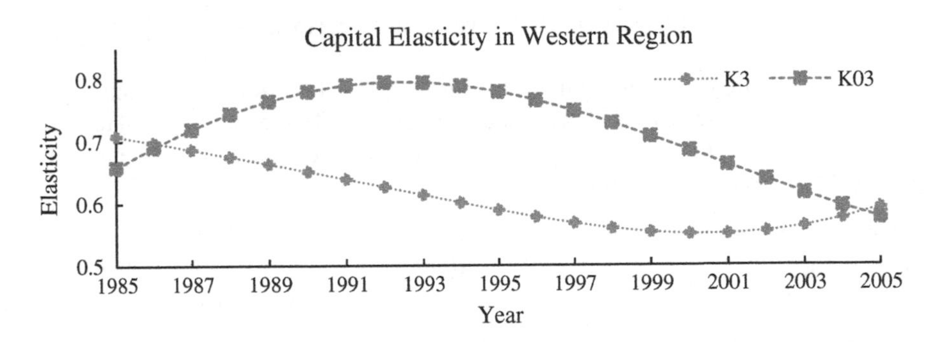

Fig. 4.4 The capital elasticity in western region

Western regions during almost all the time. Therefore, strong evidence is provided for the fact that FDI path promotes the capital elasticity in Eastern region during 1996–2005 while does not promote the capital elasticity in Middle and Western region all period.

2. **Capital Elasticity Strength: Regional Difference**

Given α_{k1} and α_{k3} the capital elasticity with FDI path model in Eastern and Western regions, α_{k01} and α_{k03} the capital elasticity with benchmark model in both regions, k' and GDP' the growth rates of capital and GDP respectively,

$$\frac{(\alpha_{k1} - \alpha_{k01})k'_{Eastern} - (\alpha_{k3} - \alpha_{k03})k'_{Western}}{GDP'_{Eastern} - GDP'_{Western}} \tag{4.30}$$

illustrates the contribution of FDI path to regional growth discrepancy in terms of the promotion of capital utilization efficiency along with time.

Expressly, calculating the average of (4.30) to time t, the FDI path contributes to regional growth discrepancy by almost 92 % between eastern and western regions during period 1995–2005, which is consistent with the conclusion by Wei (2002) that FDI's inflows lead approximately to 90 % of the gap in GDP growth rate between eastern developed regions and western undeveloped regions in China.

Figure 4.5 clearly illustrates the strengths of capital elasticity with FDI Path for Eastern, middle and western regions. The capital elasticity in Eastern region is the highest and the one in Western region is the lowest. The elasticity gap between Eastern and Middle regions gets smaller, and the capital elasticity in Middle region catches up with the one in Eastern region in 2004–2004. However, the elasticity gap between Eastern and Western regions is enlarged.

4.4.2.2 Technical Output Efficiency with FDI Path: Difference within Region

This section presents the identification of output efficiency with FDI path. The benchmark output efficiency is given by $A1 = Y_1/(K^{\alpha}L^{\beta})$. The path output efficiency with the FDI path is $A2^* = Y_2/(K^{\alpha}L^{\beta})$, where the FDI path is exogenous and assumed to be different from the factors of capital and labor. The technical

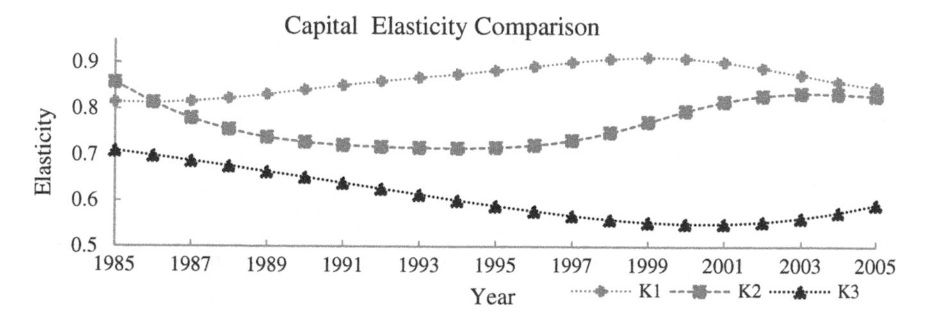

Fig. 4.5 The comparison of capital elasticity among regions

output efficiencies with FDI path in Eastern, Middle and Western regions are shown in Figs. 4.6, 4.7, 4.8.

The output efficiency has been promoted by FDI path during 1985–1995 in eastern region, the average difference reached almost 8.300 % between benchmark model and FDI path model. While output efficiency in Middle region witnesses hardly any promotion and the one in Western region suffers from a 2 % decrease.

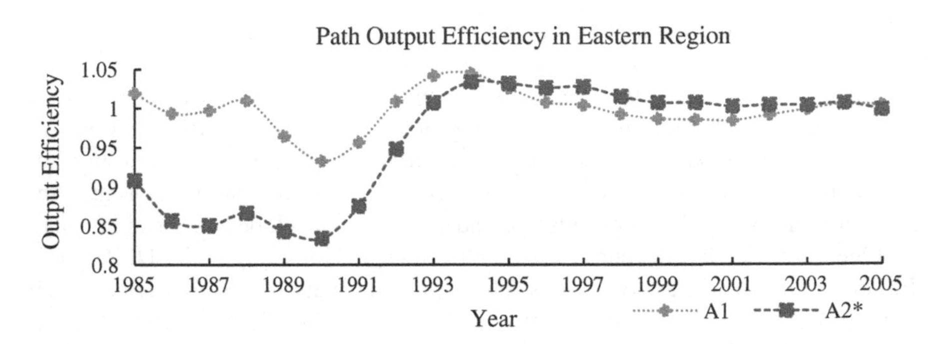

Fig. 4.6 The path output efficiency in eastern region

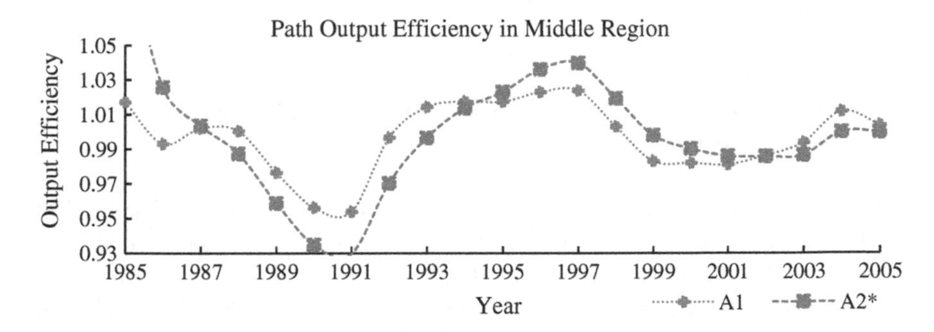

Fig. 4.7 The path output efficiency in middle region

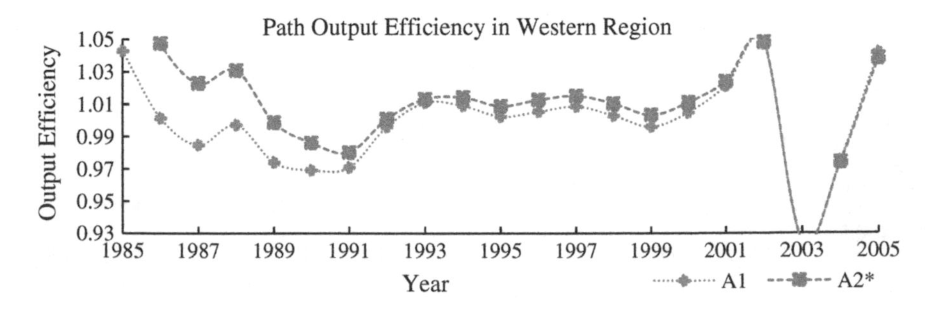

Fig. 4.8 The path output efficiency in western region

The FDI path hardly promotes the output efficiency in any region during 1996–2004. In fact, the average output efficiency has been reduced by 1.300, 0.4000 and 0.3 000 % in Eastern, Middle and Western region respectively.

In terms of regional output efficiency discrepancies, averagely, 10 and 8.300 % occur between Eastern and Western region and Eastern and Middle region during 1985–1995, which coincides with Wu (2002) conclusion.

4.5 Strategy on Crowd-in Effect of FDI Path with Urbanization Environment

FDI inflow partly depends on Chinese urbanization because of better infrastructural facilities, promising markets, and so on. In this sense, urbanization environment should be included in the FDI *Path identification* to further explore the crowd-in effect of FDI path. Based on the population agglomeration in urbanization in Chap. 3, Identification of *Crowd-in effect of FDI path with Urbanization environment* is proved as follows.

***Urbanization FDI* Path Model** S_3 is given as

$$Y_3(t) = FDI(t)^{\lambda(t)} K(t)^{\alpha(t)} L(t)^{\beta(t)} \tag{4.31}$$

The estimation of model with urbanization environment is given by the local weighed least-squares technique as below:

$$\min_{\theta(t_0)} \sum_{t=1}^{n} (\ln Y_t - (\lambda_0 + \lambda_1(t - t_0))FDI_t - (\alpha_0 + \alpha_1(t - t_0)) \ln K_t - (\beta_0 + \beta_1(t - t_0)) \ln L_t)^2 n^{-1} K_h(t - t_0) n^{-1} K_h(U - U_0)$$

$$\equiv \min_{\theta(t_0)} \sum_{t=1}^{n} (t, t_0)^2 n^{-1} K_h(t - t_0) n^{-1} K_h(U - U_0) \quad \text{(say)} \tag{4.32}$$

where U_t refers to the urbanization at time t.

Suppose that random variables $\varepsilon(t, t_0)^2$ be strong mixture sequence and $\sigma^2 = E\varepsilon(t, t_0)^2$, using the strong law of large numbers with weighted $n^{-1}K_h(t - t_0)$ by Xu and Cai (2007a, b), the following equations are proved to be true:

$$P\left[\lim_{t \to \infty} \sum_{t=1}^{n} \left(\varepsilon(t, t_0)^2 n^{-1} K_h(t - t_0) - \sigma^2\right) = 0\right] = 1 \tag{4.33}$$

And

$$P\left[\lim_{t\to\infty}\sum_{t=1}^{n}\left(\varepsilon(t,t_0)^2 n^{-1}K_h(t-t_0)n^{-1}K_h(U-U_0)-\sigma^2 n^{-1}K_h(t-t_0)\right)=0\right]=1 \tag{4.34}$$

Provide $\lim_{n\to\infty}\sum_{t=1}^{n} n^{-1}K_h(U-U_0)=\int_{-\infty}^{\infty}\frac{1}{\sqrt{2\pi}}e^{-x^2/2}dx=1$, $\lim_{n\to\infty}\sum_{t=1}^{n}(\sigma^2 n^{-1}K_h(t-t_0)-\sigma^2)=0$, and $\lim_{n\to\infty}\sum_{t=1}^{n} n^{-1}K_h(t-t_0)=\int_{-\infty}^{\infty}\frac{1}{\sqrt{2\pi}}e^{-x^2/2}dx=1$, therefore we have

Theorem 4.2 *Underlying structure S_0 can be Path identified with* ***Urbanization FDI Path*** *model S_3 based on benchmark model S_1.*

Proof

$$P\left(\lim_{t\to\infty}[Y_3(t)-Y_2(t)]\neq 0\right)$$

$$=P\left[\lim_{n\to\infty}\sum_{t=1}^{n}\left(\varepsilon(t,t_0)^2 n^{-1}K_h(t-t_0)n^{-1}K_h(U-U_0)-\varepsilon(t,t_0)^2 n^{-1}K_h(t-t_0)\right)\neq 0\right]$$

$$\leq P\left[\lim_{n\to\infty}\sum_{t=1}^{n}\left(\varepsilon(t,t_0)^2 n^{-1}K_h(t-t_0)n^{-1}K_h(U-U_0)-\sigma^2 n^{-1}K_h(t-t_0)\right)\neq 0\right]$$

$$+P\left[\lim_{n\to\infty}\sum_{t=1}^{n}\left(\varepsilon(t,t_0)^2 n^{-1}K_h(t-t_0)-\sigma^2 n^{-1}K_h(t-t_0)\right)\neq 0\right]$$

$$\leq P\left[\lim_{n\to\infty}\sum_{t=1}^{n}\left(\varepsilon(t,t_0)^2 n^{-1}K_h(t-t_0)-\sigma^2 n^{-1}K_h(t-t_0)\right)\neq 0,\lim_{n\to\infty}\sum_{t=1}^{n}(\sigma^2 n^{-1}K_h(t-t_0)-\sigma^2)=0\right]$$

$$+P\left[\lim_{n\to\infty}\sum_{t=1}^{n}(\sigma^2 n^{-1}K_h(t-t_0)-\sigma^2)\neq 0\right]$$

$$\leq P\left[\lim_{n\to\infty}\sum_{t=1}^{n}\left(\varepsilon(t,t_0)^2 n^{-1}K_h(t-t_0)-\sigma^2\right)\neq 0\right]=0$$

□

Using Eq. (4.27),

$$P\left(\lim_{t\to\infty}[Y_3(t) - Y_1(t)] \neq 0\right) = P\left(\lim_{t\to\infty}[Y_3(t) - Y_1(t)] \neq 0, \lim_{t\to\infty}[Y_1(t) - Y_2(t)] = 0\right)$$

$$+P\left(\lim_{t\to\infty}[Y_3(t) - Y_1(t)] \neq 0, \lim_{t\to\infty}[Y_1(t) - Y_2(t)] \neq 0\right)$$

$$\leq P\left(\lim_{t\to\infty}[Y_3(t) - Y_2(t)] \neq 0\right) + P\left(\lim_{t\to\infty}[Y_2(t) - Y_1(t)] \neq 0\right) = 0$$

Thus $P\left(\lim_{t\to\infty}[Y_3(t) - Y_1(t)] = 0\right) = 1$

According to ***Definition 4.2***, underlying structure S_0 can be identified with Path-Environment model S_3 based on benchmark model S_1.

4.5.1 Simulation of Crowd-in Effect of FDI Path in Middle Region

Figures 4.9, 4.10, 4.11 present the capital elasticities with urbanization environment (*Kw*) for Eastern, Middle and Western regions. With urbanization

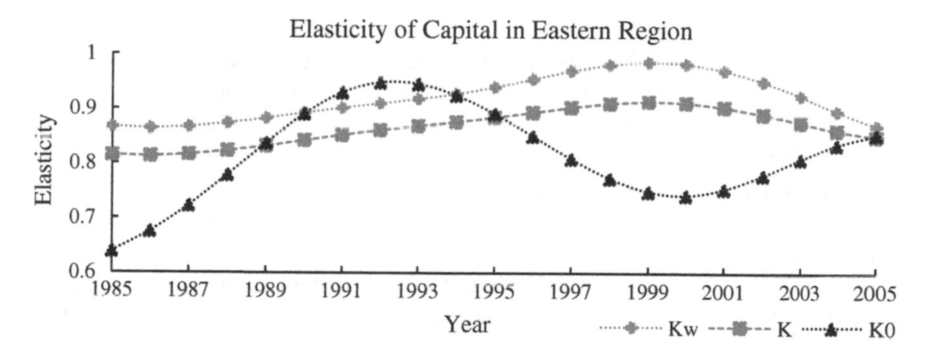

Fig. 4.9 The capital elasticity with urbanization in eastern region

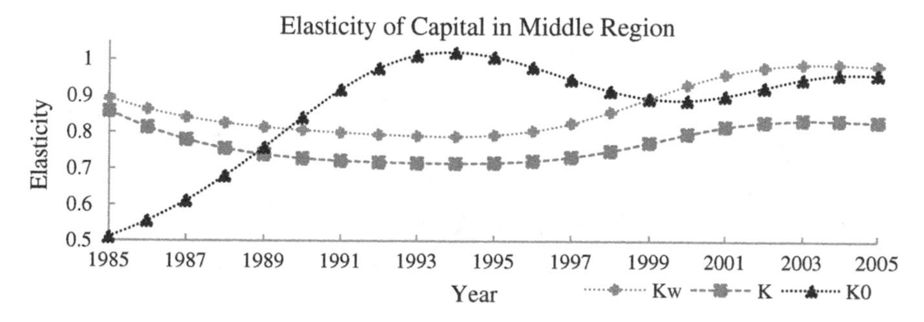

Fig. 4.10 The capital elasticity with urbanization in middle region

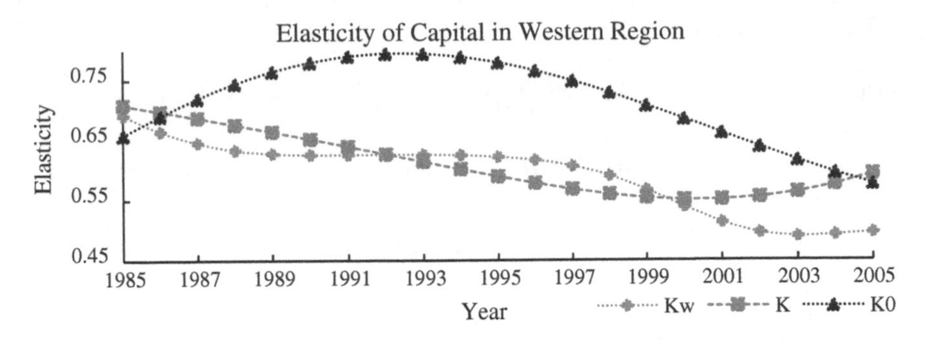

Fig. 4.11 The capital elasticity with urbanization in western region

environment considered in path-driven approach, the research finds the FDI surprisingly exerts crowd-in from crowd-out effect on capital in Middle region during 1999–2004. However, it fails in Western region.

4.5.2 Simulation of Crowd-in Effect of FDI Path in Western Region

The urbanization level 0.2500 is thought to be the critical level that renders FDI path to realize crowd-in effect of capital elasticity. Filtering the influence of left terminal point, the urbanization levels are 0.2539 in 1990 for Eastern region and 0.2468 in 1999 for Middle region, when FDI path with urbanization environment starts to exert crowd-in effect on capital for Eastern and Middle regions respectively. Some Western provinces have witnessed their urbanization levels more than 0.2500 in 1999, which will bring about crowd-in effect of FDI path at least in these provinces. The realization of crowd-in effect of FDI path on capital could be gradually obtained with increasing advanced Western provinces with high urbanization level (Figs. 4.12, 4.13).

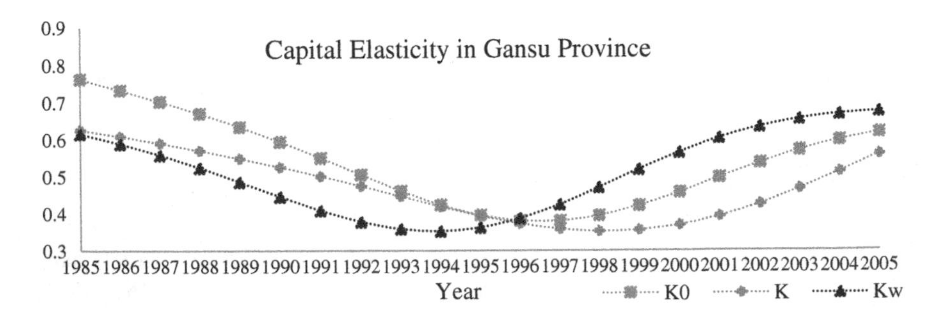

Fig. 4.12 The capital elasticity with urbanization in Gansu Province

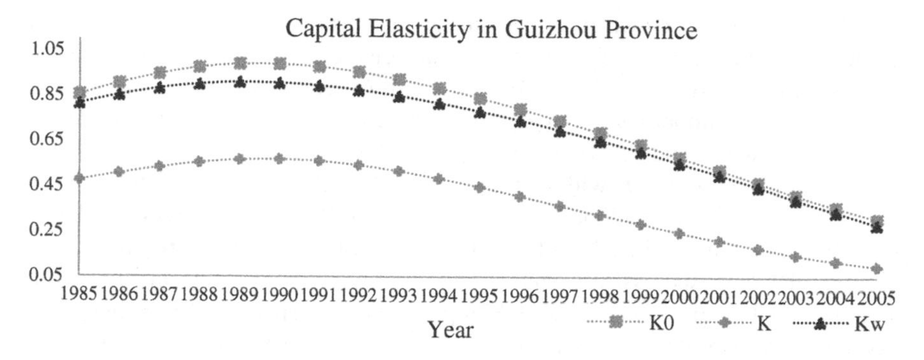

Fig. 4.13 The capital elasticity with urbanization in Guizhou Province

The urbanization levels in Gansu and Guizhou Provinces are 0.2600 and 0.1600 in 1999. Implement the strategy with path-converged design in Gansu and Guizhou provinces; FDI path in Gansu successfully realizes the crowd-in effect on capital during 1997–2005 while the one in Guizhou province still has crowd-out effect on capital during all period. The facts validate 0.2500 is the critical urbanization level that renders FDI path to realize crowd-in effect on capital in Western region.

4.6 Conclusion

The research in this chapter aims to identify and measure the changes in scale efficiency due to FDI path in China. Two contrary decision-makings are presented on whether FDI should be encouraged to reduce regional discrepancy by Wei (2002) and Wu (2002), because their researches conclude that FDI explain about 20 and 90 % of regional discrepancy between Eastern and Western China respectively. To assist the decision makers, a new Path converged approach is developed with application to identify how FDI affects regional productivity discrepancy in China, and then to design reasonable decision makings for reducing the discrepancy.

Methodologically, path driven presents an alternative approach to solve three problems by Fine (2000) since Path or Environment Model fixes on the framework of the generalized CD production function when accession of new factor, achieves choice of models; the relativity between Path or Environment Model and Benchmark model avoids independency hypothesis among factors; and time-varying coefficient depicts a persistent and varied impact of factors shocks over time.

The FDI *Path* identification discovers that the regional productivity discrepancy due to FDI Path mainly attributes to capital elasticity rather than output efficiency. It successfully unifies the diverse conclusions of Wei (2002) and Wu (2002) via time-varying capital elasticity and output efficiency respectively instead of constant capital elasticity and output efficiency in their studies.

FDI path exerts a crowd-in effect on capital elasticity after 1996 in Eastern region and a crowd-out effect on capital elasticity in both Middle and Western regions during almost all the time. With the aim of reducing the regional growth discrepancy, it is important to change the crowd-out effect to crowd-in effect of FDI path in Middle and Western regions.

FDI *Path-Environment* with urbanization is further developed to explore strategy on narrowing regional discrepancy when Path-Environment Model is proved to converge to benchmark model. With urbanization environment considered, FDI path successfully changes to the crowd-in effect from crowd-out effect in Middle region. Furthermore, FDI path can exert crowd-in effect on capital elasticity in Western provinces with their urbanization levels higher than 0.2500.

References

Chen, L. (2007) Semiparametric identification of structural dynamic optimal stopping time models. CeMMAP Working Paper, CWP06/07.

Chen, X., Hu, Y., & Arthur, L. (2007) Nonparametric identification of regression models containing a misclassified dichotomous regressor without instruments. CeMMAP Working Paper.

Fedderke, J. W., & Romm, A. T. (2006). Growth impact and determinants of foreign direct investment into South Africa, 1956–2003. *Economic Modelling, 23*, 738–760.

Fine, B. (2000). Endogenous growth theory: A critical assessment. *Cambridge Journal of Economics, 24*, 245–264.

Goss, E., Wingender, J. R., & Torau, J. M. (2007). The contribution of foreign capital to U.S. productivity growth. *The Quarterly Review of Economics and Finance, 47*(3), 383–396.

Hu, A. G. (2007). Technology parks and regional economic growth in China. *Research Policy, 36*, 76–87.

Lai, M., Peng, S., & Bao, Q. (2006). Technology spillovers, absorptive capacity and economic growth. *China Economic Review, 17*, 300–320.

Li, X., & Liu, X. (2005). Foreign direct investment and economic growth: An increasingly endogenous relationship. *World Development, 33*, 393–407.

Liu, Z. (2008). Foreign direct investment and technology spillovers: Theory and evidence. *Journal of Development Economics, 85*, 176–1938.

Luo, C. (2007). FDI, domestic capital and economic growth: Evidence from panel data at China's provincial level. *Frontiers of Economics in China, 2*(1), 92–113.

Ma, Y. (2006). The effect of FDI on China's economic growth. *Statistical Research, 3*, 51–54.

Ma, J., & Guo, W. (2007). The impact of agglomeration economy on FDI absorption in China. *Contemporary Finance and Economics, 8*, 109–112.

Mahajan, A. (2006). Identification and estimation of regression models with misclassification. *Econometrica, 74*(3), 631–664.

Ng, L. F., & Tuan, C. (2006). Spatial agglomeration, FDI, and regional growth in China: Locality of local and foreign manufacturing investments. *Journal of Asian Economics, 17*, 691–713.

Qian, X. Y. (2007). Do regional economic gaps result from FDI? The forming of secondary statistical error. *Statistical Research, 24*, 83–87.

Ramirez, M. D. (2006). Is Foreign Direct Investment Beneficial for Mexico? An Empirical Analysis, 1960–2001. *World Development, 34*, 802–817.

Ran, J., Voon, J. P., & Li, G. (2007). How does FDI affect China? Evidence from industries and provinces. *Journal of Comparative Economics, 35*, 774–799.

Schennach, S. M., Hu, Y. & Lewbel, A. (2007) Nonparametric identification of the classical errors-in-variables model without side information. CeMMAP Working Paper, CWP14/07.

Stone, C. J. (1984). An asymptotically optimal window selection rule for kernel density estimates. *The Annals of Statistics, 12*, 1285–1297.

Wei, H. K. (2002). Effects of foreign direct investment on regional economic growth in China. *Journal of Economic Research, 4*, 19–26.

Wu, J. (2002). Regional discrepancy of FDI in China and its effect on economic growth. *Journal of Economic Research, 4*, 27–34.

Xin, L., & Deng, S. (2007). On crowding-in effect and crowding-out effect of FDI to private investment. Modern Finance and Economics, 27, 9–14.

Xu, B., & Cai, G. (2007a). Some maximal inequalities and complete convergences of non-identically distributed negatively associated random sequences. *Journal of Interdisciplinary Mathematics, 10*(3), 327–338.

Xu, B., & Cai, G. (2007b). Maximal inequality and complete convergences of non-identically distributed negatively associated sequences. *Applied Mathematics A: Journal of Chinese Universities Series B, 22*(3), 316–324.

Xu, B., & Watada, J. (2007b) A Time-varying Elasticity Production Function Model with Urbanization Endogenous Growth: Evidence of China, In *Proceedings of Second International Conference on Innovative Computing, Information and Control, 2007, ICICIC'07. 0-7695-2882-1/07* (pp. 375-375).

Zhang, J. (2008). Estimation of China's provincial capital stock (1952–2004) with applications. *Journal of Chinese Economic and Business Studies, 6*(2), 177–196.

Chapter 5
Changes in Technical Efficiency of FDI

It is well-accepted that the absorption of FDI will stimulate economic growth through technical improvement. In this sense, the changes in technical efficiency would be important in identifying the changes in production efficiency. Similar to Chap. 4, this chapter will identify and measure the second aspect of indirect change in production efficiency, which refers to how FDI absorption brings about change in technical efficiency and its decompositions in production system.

5.1 Introduction

Since Farrell (1957) pioneering research, most of previous researches have been presented to compute and analyze technical efficiency from economics and statistics perspectives. However, none of any application-based research to realize technical efficiency from management engineering perspective was found in literatures because the production frontier surface, the standard method in existing measurements of DEA (Charnes et al. 1979; Lee 2008; Cook and Seiford 2009) and SFA (Meeusen and Broeck 1977; Førsund et al. 1980; Schmidt 1985, Lovell and Schmidt 1988; Bauer 1990; Greene 1993; Kumbhakar and Lovell 2000; Cornwell and Schmidt 2008), can not be observed and it is hard to control the variables and isolate the relations to identify an unknown underlying production function.

The measure of technical efficiency is usually referred to as Total Factor Productivity (TFP) in "Solow" residual model. With previous research framework of DEA and SFA, the productivity growth is decomposed into technical progress and efficiency improvement. Whether the productivity growth attributes to technical progress or efficiency improvement in a region or industry has been the focus of both theoretical researches and governmental decision-makers.

Murakami (2007) demonstrates the entry of foreign-owned firms has a positive effect on the productivity of local firms in Japan as a result of technology spillovers in the long run. Fare et al. (2001) find that productivity growth is generally

B. Xu et al., *Changes in Production Efficiency in China*,
DOI: 10.1007/978-1-4614-7720-4_5,

achieved through technical progress, and the efficiency change negatively contributes to productivity growth for Taiwanese manufacturing.

However, Cook and Uchida (2002) find that efficiency improvement dominates technical progress in developing countries. Lam and Shiu (2008) point that the differences in efficiency scores are mainly due to the differences in operating environments of different provinces, rather than the efficiency performance of telecommunications enterprises. Kim and Park (2006) show both domestic and foreign R&D played an important role in increasing efficiency and technical progress in Korean manufacturing. However, domestic R&D has more effect on technical progress, while foreign R&D has played a relatively stronger role in efficiency improvement.

Consequently, various conclusions on technical efficiency confuse the decisions on technology innovation. Kim and Han (2001) point if the low production efficiency results from weak technical progress, then a decision that will induce technological innovation should be recommended. If high rates of technical progress coexist with deteriorating technical efficiency, resulting in slow production efficiency growth, then a decision to increase the efficiency with which a known technology is applied is required, which might include improvements in learning-by-doing processes and in managerial practices.

The diverse conclusion of previous researches definitely confuses the decision-making. However, the realization of technical efficiency is of crucial importance for both decision makers and theoretical applications. When a new factor enters into a production system, the identification of relative change of technical efficiency due to a new factor is much more meaningful than the identification of technical efficiency itself, which also illuminates the possible realization strategy of technical efficiency.

Taken a deep thinking of growth in China, on the one hand, the history of development in China is a history of regional development. On the other hand, special incentive policies have been offered in China to attract foreign direct investment (FDI). Despite of the relative small amount of FDI compared to the native capital, it exerts great role on growth (Lai et al. 2006). More specifically, Luo (2007) points out FDI have both direct and indirect effects on economic growth, of which the direct effects is of insignificance, and that the indirect impacts are obtained through improving technical efficiency in China. Moreover, FDI is thought to be related to regional growth in China (Kuo and Yang 2008; Ma and Guo 2007; Ma 2006). Since FDI is not equivalent to native capital and labor in economic system, and it impacts on growth with promotion of productivity through technological spillovers (Mastromarco and Ghosh 2009; Bitzer and Kerekes 2008; Liu 2008), it is introduced as a new factor in production system.

In this sense, the research establishes an innovative Path-converged design technique not only to identify but also to realize change of technical efficiency in different regions of China. The research takes FDI as an underlying path in path-converged design to explore the efficiency change among regions in China. More specifically, this research includes the ***Path-converged*** design based on underlying structure, benchmark model and path model with application to firstly decompose

TFP into technical level and efficiency level, to secondly identify whether productivity growth attributes to ***technical progress*** or ***efficiency improvement*** with introduction of FDI in different regions of China, and to thirdly realize technical efficiency in less developed regions, which is used to validate the feasibility of path-converged design technique.

The rest of this chapter is organized as follows. Section 5.2 presents the ***Path-Converged*** technique with technical level, efficiency level and total factor productivity indexes. Section 5.3 identifies ***technical progress*** and ***efficiency improvement*** with FDI Path among regions. Section 5.4 measures the realization strategy of technical progress in both Middle and Western regions. And Sect. 5.5 finally concludes.

5.2 Path-Converged Technique

5.2.1 FDI-Path Approach

Assume the well-known C-D production function as equilibrium of economic system, which is presented by the equilibrium structure S_0:

$$Y(t) - A_0(t)K(t)^{\alpha}L(t)^{\beta} = 0 \tag{5.1}$$

where K and L denotes the inputs of labor and capital, α and β are given in (5.11) below and correspond to output elasticity of capital and labor respectively, A_0 refers to the technical level of production.

Since the initial equilibrium structure S_0, taken as the underlying structure, is unknown, the first step for further research is to estimate S_0, which is obtained as follows.

Define a benchmark model S_2 by:

$$Y_2(t) - K(t)^{\alpha(t)}L(t)^{\beta(t)} = 0 \tag{5.2}$$

Intuitively, taken the view of statistics, the following assumption needs to be satisfied for study:

$$P\left(\lim_{t\to\infty}[Y_2(t) - Y(t)] = 0\right) = 1 \tag{5.3}$$

Assumption (5.3) indicates that $Y_2(t)$ is a strong consistent estimator of $Y(t)$.

Now, suppose a new exterior factor FDI enters into the initial equilibrium structure S_0, it is taken as FDI path in the study. To depict the change process, define the path model S_1 by:

$$Y_1(t) - FDI(t)^{\lambda(t)}K(t)^{\alpha(t)}L(t)^{\beta(t)} = 0 \tag{5.4}$$

The change process from equilibrium production structure S_0 to the disequilibrium production model S_1 is represented as follows:

$$Y(t) - FDI(t)^{\lambda(t)} K(t)^{\alpha(t)} L(t)^{\beta(t)} \neq 0 \text{ if } FDI \neq \equiv 1 \tag{5.5}$$

Since FDI is a new factor that introduced in the initial equilibrium, the change process from equilibrium to disequilibrium structure also presents reallocation efficiency of the new factor FDI.

Suppose $FDI(t) = 1$ which return to (5.2) denotes static equilibrium, then $FDI(t)$ in (5.4) is an indicator of the deviation degree of dynamic disequilibrium from static equilibrium in (5.2). Coefficient $\lambda(t)$ represents the strength of disequilibrium.

The final process, after taking some time to adjust the disequilibrium (5.5) into re-equilibrium of resource allocation (5.2), involved is how to model the changing process of the production structure *from the disequilibrium to equilibrium*. To reach the goal, ***Path convergence design*** of the new resource factor is need.

$$P\left(\lim_{t\to\infty}[Y_1(t) - Y_2(t)] = 0\right) = 1 \tag{5.6}$$

In the change process from equilibrium to disequilibrium, and disequilibrium to equilibrium production structure, it achieves identification of the change effects of underlying production structure generated by the path model converge to benchmark model along with $P(t)$ path.

Inspired by the approaches of DEA and SFA with technical levels, the objectives of this research are to first identify the change effects of technical and efficiency levels when exterior FDI accesses into the production structure.

Suppose t is at a neighborhood of t_0, the local approximations are given as:

$$\begin{aligned} \lambda(\mathrm{t}) &= \lambda(\mathrm{t}_0) + \lambda\prime(\mathrm{t})(\mathrm{t}-\mathrm{t}_0) + \mathrm{o}(\mathrm{t}-\mathrm{t}_0) \equiv \lambda_0 + \lambda_1(\mathrm{t}-\mathrm{t}_0) + \mathrm{o}(\mathrm{t}-\mathrm{t}_0) \\ \alpha(\mathrm{t}) &= \alpha(\mathrm{t}_0) + \alpha\prime(\mathrm{t})(\mathrm{t}-\mathrm{t}_0) + \mathrm{o}(\mathrm{t}-\mathrm{t}_0) \equiv \alpha_0 + \alpha_1(\mathrm{t}-\mathrm{t}_0) + \mathrm{o}(\mathrm{t}-\mathrm{t}_0) \\ \beta(\mathrm{t}) &= \beta(\mathrm{t}_0) + \beta\prime(\mathrm{t})(\mathrm{t}-\mathrm{t}_0) + \mathrm{o}(\mathrm{t}-\mathrm{t}_0) \equiv \beta_0 + \beta_1(\mathrm{t}-\mathrm{t}_0) + \mathrm{o}(\mathrm{t}-\mathrm{t}_0) \end{aligned} \tag{5.7}$$

where $\mathrm{x}'(\mathrm{t}) = \mathrm{dx/dt}$ is the time derivatives of corresponding variable. And, if $\mathrm{o}(\mathrm{t} - \mathrm{t}_0) \to 0$, if $\mathrm{t} \to \mathrm{t}_0$.

Thus Eqs. (5.2) and (5.4) can be respectively expressed by:

$$\begin{aligned} Y_1(t) &= K(t)^{\alpha(t)} L(t)^{\beta(t)} \\ &= K(t)^{\alpha_1(t)} L(t)^{\beta_1(t)} K(t)^{\alpha_0-\alpha_1(t_0)} L(t)^{\beta_0-\beta_1(t_0)} (K(t)L(t))^{o(t)} \\ &= A_1 K(t)^{\alpha} L(t)^{\beta} (K(t)L(t))^{o(t-t_0)} \end{aligned} \tag{5.8}$$

and

$$\begin{aligned}Y_2(t) &= FDI(t)^{\lambda(t)}K(t)^{\alpha(t)}L(t)^{\beta(t)}\\ &= FDI(t)^{\lambda_1(t)}K(t)^{\alpha_1(t)}L(t)^{\beta_1(t)}FDI(t)^{\lambda_0-\lambda_1(t_0)}K(t)^{\alpha_0-\alpha_1(t_0)}L(t)^{\beta_0-\beta_1(t_0)}(U(t)K(t)L(t))^{o(t-t_0)}\\ &= A_2FDI(t)^{\lambda}K(t)^{\alpha}L(t)^{\beta}(FDI(t)K(t)L(t))^{o(t-t_0)}\end{aligned} \tag{5.9}$$

where

$$A_1(t) = K(t)^{\alpha_1(t)}L(t)^{\beta_1(t)},\ A_2(t) = FDI(t)^{\lambda_1(t)}K(t)^{\alpha_1(t)}L(t)^{\beta_1(t)} \tag{5.10}$$

$$\alpha = \alpha_0 - \alpha_1 t_0,\ \beta = \beta_0 - \beta_1 t_0,\ \lambda = \lambda_0 - \lambda_1 t_0 \tag{5.11}$$

A_1 and A_2 reflect production technical level of Benchmark and Path model respectively.

5.2.2 *Technical Level, Efficiency Level and Total Factor Productivity*

Rewriting Eqs. (5.2) and (5.4), it is shown that

$$\begin{aligned}Y_1(t) &- K(t)^{\alpha(t)}L(t)^{\beta(t)}\\ &= K(t)^{\alpha_1 t-\alpha_1 t_0}L(t)^{\beta_1(t)-\beta_1 t_0}(K(t)L(t))^{o(t)}K(t)^{\alpha_0}L(t)^{\beta_0}\\ &= A_1(t,t_0)K(t)^{\alpha_0}L(t)^{\beta_0}\end{aligned} \tag{5.20}$$

where α_0 and β_0 are taken as capital and labor elasticity respectively, $A_1(t,t_0) = K(t)^{\alpha_1 t-\alpha_1 t_0}L(t)^{\beta_1(t)-\beta_1 t_0}(K(t)L(t))^{o(t-t_0)}$ is the total factor productivity.

Taking $A_1(t,t_0) = exp(-u_t + v_t)$, or $(K(t)L(t))^{o(t-t_0)} = exp(-u_t + v_t)$ $\left(K(t)^{\alpha_1 t_0-\alpha_1 t}L(t)\right)^{\beta_1 t_0-\beta_1 t}$

where v is a random variable and u is a measure of inefficiency, *which is stochastic frontier production model* (Battese and Coelli 1995), where the distribution of u is taken to be the non-negative truncation of t the distribution is often assumed to be half-normal, truncated normal, gamma or exponential, and estimate technical efficiency using maximum likelihood estimation (MLE).

$$\begin{aligned}Y_2(t) &= FDI(t)^{\lambda(t)}K(t)^{\alpha(t)}L(t)^{\beta(t)}\\ &= FDI(t)^{\lambda_1 t-\lambda_1 t_0}K(t)^{\alpha_1 t-\alpha_1 t_0}L(t)^{\beta_1 t-\beta_1 t_0}(FDI(t)K(t)L(t))^{o(t-t_0)}FDI(t)^{\lambda_0}K(t)^{\alpha_0}L(t)^{\beta_0}\\ &= A_2(t,t_0)FDI(t)^{\lambda_0}K(t)^{\alpha_0}L(t)^{\beta_0}\end{aligned} \tag{5.21}$$

where λ_0 is output elasticity of FDI. The total factor productivity here is given by

$$A_2(t,t_0) = FDI(t)^{\lambda_1 t-\lambda_1 t_0} K(t)^{\alpha_1 t-\alpha_1 t_0} L(t)^{\beta_1 t-\beta_1 t_0} (FDI(t)K(t)L(t))^{o(t-t_0)}$$

The A_2 can be decomposed into two economically meaningful sources of growth by

$$FDI(t)^{\lambda_1 t-\lambda_1 t_0} K(t)^{\alpha_1 t-\alpha_1 t_0} L(t)^{\beta_1 t-\beta_1 t_0} \tag{5.22}$$

$$(FDI(t)K(t)L(t))^{o(t-t_0)} \tag{5.23}$$

The time in (5.22) varies from t_0 to t_1 while the inputs of K and L keep fixed on time t. In this sense, Eq. (5.22) expresses that the output growth is attributable to the technical change along with time rather than the increase in inputs.

Equation (5.23) illustrates that the output growth changes not only because of the variation of time but also the inputs variation in inputs at time t in the perspective of efficiency level. Intuitively, the efficiency equals to be 1 when time goes to infinity.

Follow to the method in Malmquist index of DEA approach on TFP growth, the technical change of TFP growth during period t to t_0 is defined by:

$$TC = \left(FDI(t)^{\lambda_1 t-\lambda_1 t_0} K(t)^{\alpha_1 t-\alpha_1 t_0} L(t)^{\beta_1 t-\beta_1 t_0} FDI(t_0)^{\lambda_1 t-\lambda_1 t_0} K(t_0)^{\alpha_1 t-\alpha_1 t_0} L(t_0)^{\beta_1 t-\beta_1 t_0}\right)^{1/2} \tag{5.24}$$

TC is the geometric average of the ratios of both the invariable factor inputs and the varying time. TC measures the technical change between periods t and t_0: the technical declines if TC < 1, remains unchanged if TC $= 1$ and witnesses progress if TC > 1.

And the efficiency change is defined as:

$$EC = (FDI(t)K(t)L(t))^{o(t-t_0)} \Big/ (FDI(t_0)K(t_0)L(t_0))^{o(t-t_0)} \tag{5.25}$$

EC measures the magnitude of efficiency change between periods t and t_0. EC > 1, EC $= 1$ and EC < 1 indicate the efficiency improves, remains the unchanged, and declines respectively.

To sum up, the technical change and efficiency level of TFP of Model (5.2) are estimated as follows. On the one hand, after the estimation of α_0, α_1, β_0, β_1, λ_0, and λ_1 by model (5.11), the technical change at time t based on time t_0 is:

$$\left(FDI(t)^{\lambda_1 t-\lambda_1 t_0} K(t)^{\alpha_1 t-\alpha_1 t_0} L(t)^{\beta_1 t-\beta_1 t_0} FDI(t_0)^{\lambda_1 t-\lambda_1 t_0} K(t_0)^{\alpha_1 t-\alpha_1 t_0} L(t_0)^{\beta_1 t-\beta_1 t_0}\right)^{1/2}$$

And the technical level is given by:

$$Tech(t) = FDI^{\lambda_1 t}(t) K^{\alpha t}(t) L^{\beta_1 t}(t) \tag{5.26}$$

On the other hand, the efficiency level $(FDI(t)K(t)L(t))^{o(t-t_0)}$ at point t when $t = t_0$ in Eq. (5.23) is given by:

$$EFF(t) = Y(t)\left(FDI(t)^{\lambda_0}K(t)^{\alpha_0}L(t)^{\beta_0}\right)^{-1} \quad (5.27)$$

And the efficiency change at point t to t_0 is obtained by the difference of estimator (5.27).

If the efficiency level equals to 1, then the total factor productivity turns out to be the technical change. In other words, technical inefficiency does not exist, the above decomposition implies that technical inefficiency does not affect Total Factor Productivity growth, as in the Solow residual approach.

5.3 Technical Progress and Efficiency Improvement

The data used in this chapter is the same data in Chap. 4. Given Models (5.2) and (5.4) estimated by the local weighed least-squares technique, this section provides the empirical decomposition of total factor productivity into technical progress and efficiency improvement with FDI Path identification through productivity-oriented growth among regions in China. The empirical results can be split into three parts, the first part gives the technical level, the second part illustrates the efficiency level, and the third part presents technical change and efficiency change for each region respectively.

5.3.1 Increasing or Descending Trend of Technical Level

The technical level is given by: $\hat{T}ech(t) = FDI^{\hat{\lambda}_1 t}(t)K^{\hat{\alpha}_1 t}(t)L^{\hat{\beta}_1 t}(t)$.

Figure 5.1 illustrates the technical levels of benchmark and FDI-path models in eastern region. An exciting fact is that the trend of technical level is raised, which reverses the descending trend of the technical level after 1994 in Eastern region. It presents that the technical level experiences a negative impact of FDI path based on the one of benchmark model before 2004. Fortunately, the technical level after 2004 witnesses a positive impact of FDI path with the one of benchmark.

Figures 5.2 and 5.3 give the illustrations of technical levels of benchmark and FDI path models in Middle and Western region. The technical levels in both regions all witness positive impact of FDI path compared with the ones of benchmark after 1991, however, both of them experience a declining trend and the levels in both regions fall rapidly, with the one in middle region nearly close to the one in Western region almost the same to their respective benchmark level in year 2004. The fact reflects that FDI path yields only adequate result and less fortunate result in promoting technical levels for Middle and Western regions despite of the slight positive spillover in both regions.

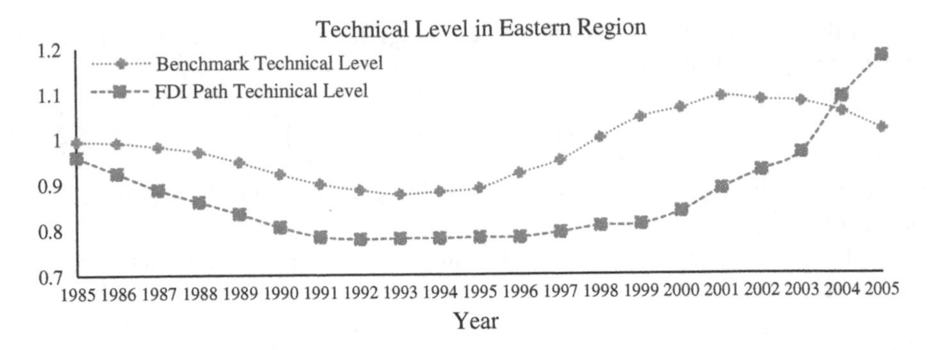

Fig. 5.1 Technical levels of benchmark and FDI path models in Eastern region

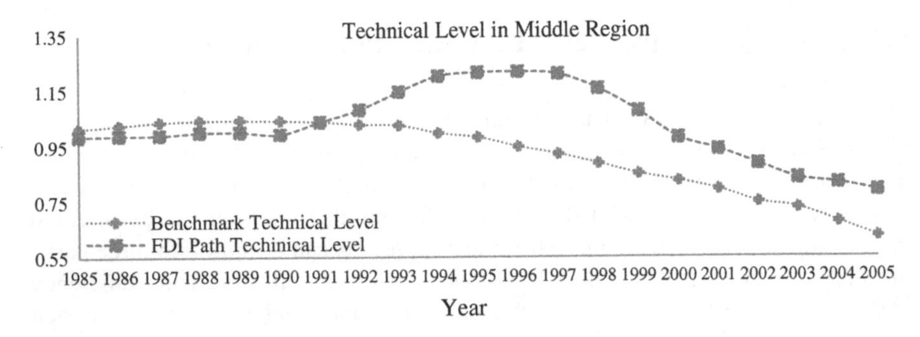

Fig. 5.2 Technical levels of benchmark and FDI path models in Middle region

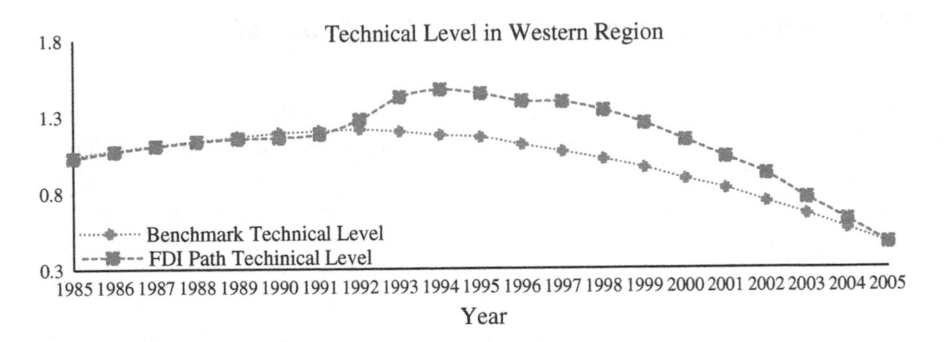

Fig. 5.3 Technical levels of benchmark and FDI path models in Western region

5.3.2 Positive or Negative Impact of Efficiency Level

Besides the exploration of technical level, which is the capability of technological innovation, the efficiency level is another issue of public interest, which is the capability of technological application. Efficiency level is presented by Eq. (5.27).

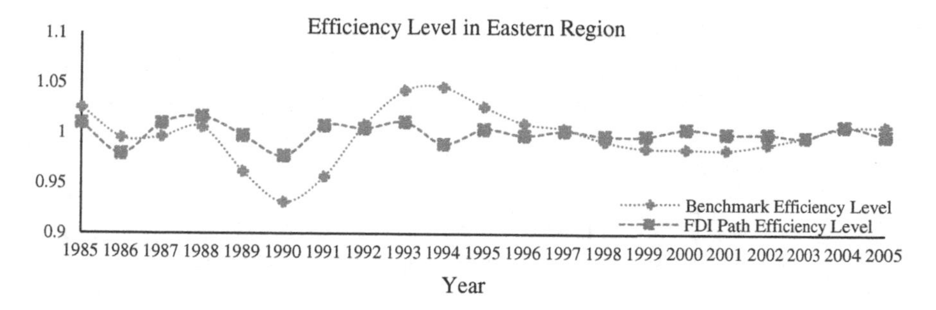

Fig. 5.4 Efficiency levels of benchmark and FDI path models in Eastern region

Figures 5.4 and 5.5 illustrate the efficiency levels of benchmark and FDI path models in Eastern region. The efficiency level witnesses slight positive impact with FDI path compared with the one of benchmark during almost all period despite of a slight negative impact during 1993–1998 in Eastern region and Middle region. The efficiency level witnesses slight negative impact with FDI path compared with the one of benchmark during almost all after 1993 in Western region in Fig. 5.6. The efficiency level does not witness significant improvement since no substantial differences are observed between the efficiency level with FDI path and the one with benchmark in each region.

5.3.3 Technical and Efficiency Changes

Besides the illustrations of technical level and efficiency levels, the rest focus is on the change effects of both technical and efficiency levels in different regions. The technical change is given by Eq. (5.24) and efficiency change is given by Eq. (5.25). The $t_0 = 2003$ in Eastern region since 2003 is the nearest year to the intersection of technical level with benchmark and the level with FDI path.

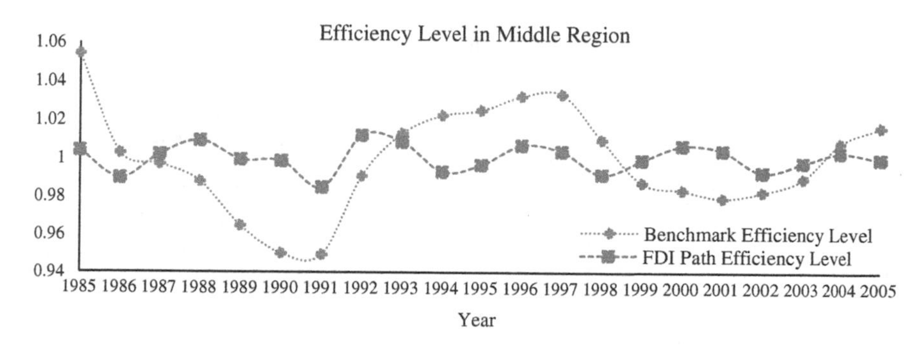

Fig. 5.5 Efficiency levels of benchmark and FDI path models in Middle region

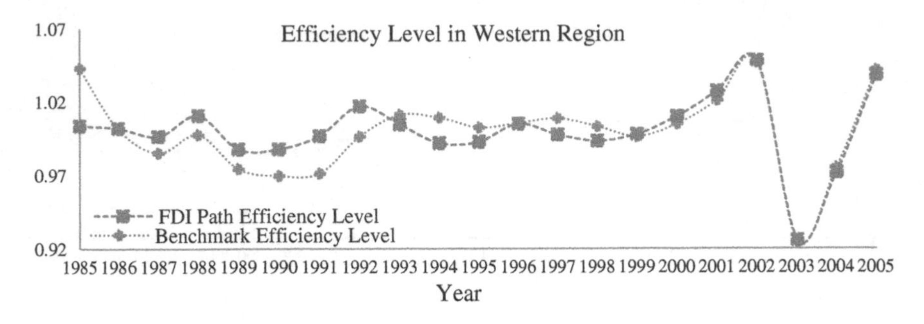

Fig. 5.6 Efficiency levels of benchmark and FDI path models in Western region

Figure 5.7 gives the change trends of technical levels of benchmark and FDI path models in Eastern region. The technical change of FDI path is higher than the one of benchmark during 1985–2001, and significant difference is observed between the two changes before 1993. The technical change of FDI path experiences a declining trend and the one of benchmark first goes up during 1985–1992 and slowly declines during 1993–2005, that is why their difference becomes smaller and smaller and they finally coincide with each other during 2002–2004.

Figure 5.8 illustrates the efficiency changes of benchmark and FDI path models in Eastern region. The efficiency change with FDI path model tends to be more stable than the one with benchmark model, and the efficiency change with FDI path model is larger than the one with benchmark model during almost all period except 1993–1995. Both facts similarly express that FDI absorption could raise the efficiency level much more.

Figure 5.9 explains the technical changes of benchmark and FDI path models in Middle region. Similar to the situation in Eastern region, the technical change of FDI path is greater than the one of benchmark. A relative large difference is observed between the two changes during 1993–2002 compared with other periods. Declining trend is observed for the technical change with FDI path during 1992–2004. Similarly, the efficiency change of FDI path model is larger than the

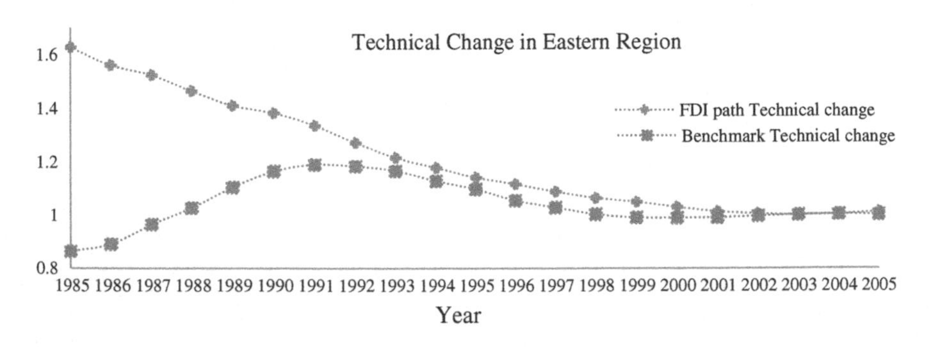

Fig. 5.7 Technical changes of benchmark and FDI path models in Eastern region

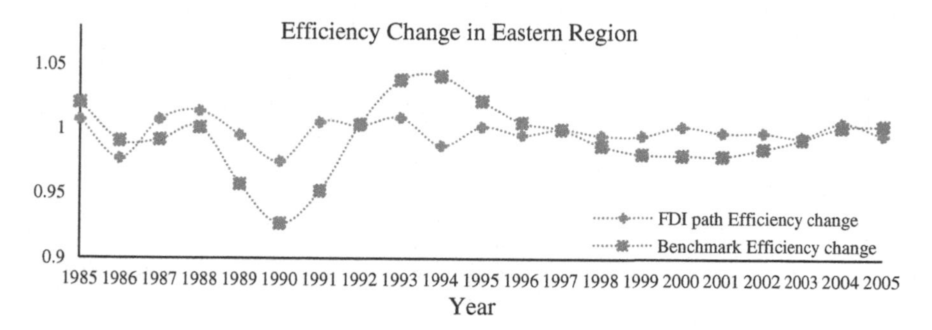

Fig. 5.8 Efficiency changes of benchmark and FDI path models in Eastern region

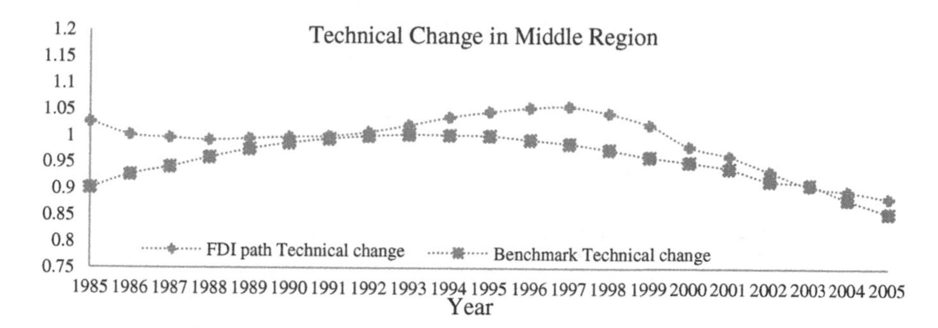

Fig. 5.9 Technical changes of benchmark and FDI path models in Middle region

one of benchmark during almost all period, with little fluctuations during period 1994–1998 (Fig. 5.10).

Figures 5.11 and 5.12 present the technical changes and efficiency changes of benchmark and FDI-path models in Western region. Different from the situations in both Eastern and Middle regions, no significant differences are observed between the technical changes and efficiency changes with both models.

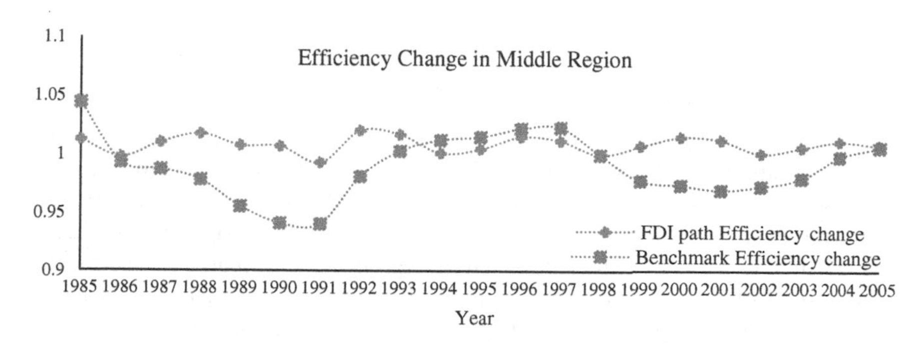

Fig. 5.10 Efficiency changes of benchmark and FDI path models in Middle region

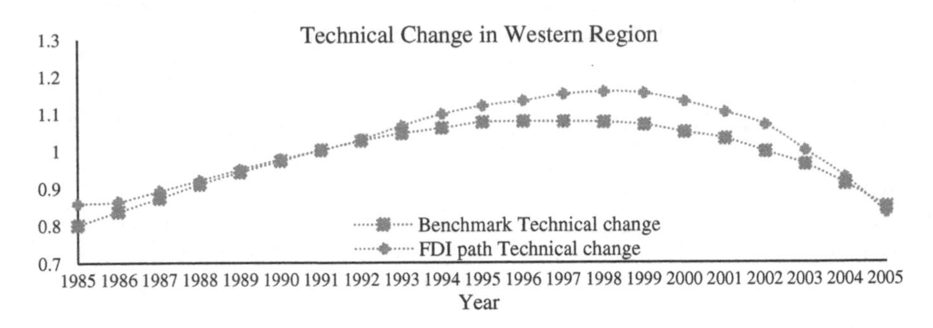

Fig. 5.11 Technical changes of benchmark and FDI path models in Western region

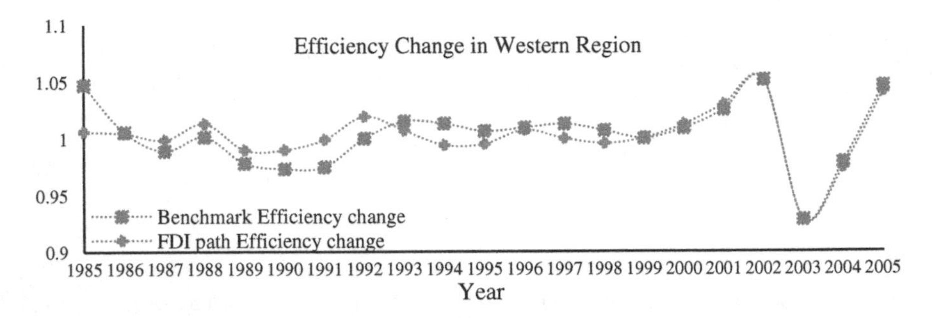

Fig. 5.12 Efficiency changes of benchmark and FDI path models in Western region

To sum up, the technical change with FDI path is larger than the one of benchmark in each region, implying the technical level fluctuation with FDI path is greater than the one with benchmark and the introduce of FDI path will bring about technology progress. Moreover, the difference of technical changes tends to greater than the difference of efficiency changes between the FDI path model and benchmark model in each region, which suggests the production efficiency mainly attributes to technical progress rather than efficiency improvement in both Eastern and Middle regions.

5.4 Realization Strategy of Technical Progress in Middle and Western Regions

The identification illustrates that productivity growth mainly attributes to technical progress rather than efficiency improvement with FDI path in both Eastern and Middle regions. Moreover, the technical level increases in Eastern region while the ones decline in both Middle and Western regions. The research concerns the varying trend of technical level rather than the absolute level of technical level, because the increasing trend of technical level will successfully avoids the diminishing marginal returns of capital and labor inputs.

To pursue productivity growth in Middle and Western regions, the realization strategy should change declining technical level to an increasing one to avoid diminishing marginal returns. The research can realize technical progress in both regions with path-converged design technique.

5.4.1 Realization Strategy in Middle Region

Figure 5.2 illustrates the technical levels in Middle region experience declining trend during period 1996–2005. GDP per capita, as an indicator of technical efficiency, is about 0.8 times of the national level in 1996 in Middle region. And the times of GDP per capita decline successively in 1996–2005. In these sense, the 0.8000 times is the necessary level that sustains the increasing technical level. Several Middle provinces have presented their GDP per capita more than 0.8000 times of national level in 1996. The realization of increasing technical level may be obtained at least in these advanced provinces in Middle region.

The GDP per capita in Shanxi (山西) and Anhui Provinces are 0.8000 and 0.6800 times of national GDP per capita in 1996. Implement the strategy with path-converged design in Shanxi and Anhui provinces; the technical level in Anhui province witnesses a more violent decline and the technical level in Shanxi province successfully realizes steady increase. The facts validate the 0.8000 times of national GDP per capita is a critical GDP per capita for Middle region to sustain its increasing technical level and to avoid the decreasing marginal returns (Figs. 5.13, 5.14).

5.4.2 Realization Strategy in Western Region

Similarly, the technical level in Western region declines successively during 1994–2005. The GDP per capita is about 0.6500 times of the national level in 1994

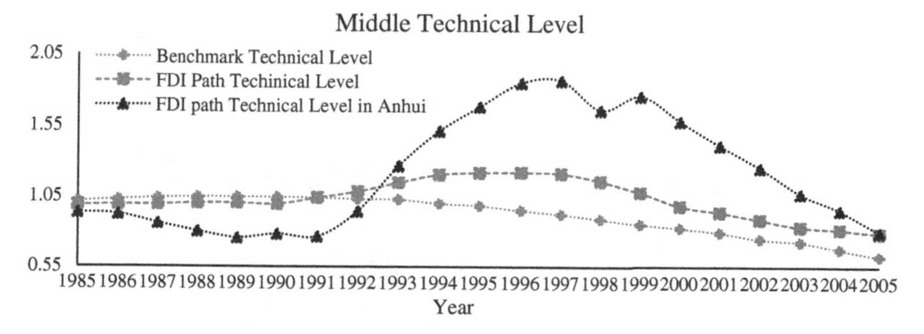

Fig. 5.13 Simulation of technical efficiency in Anhui province

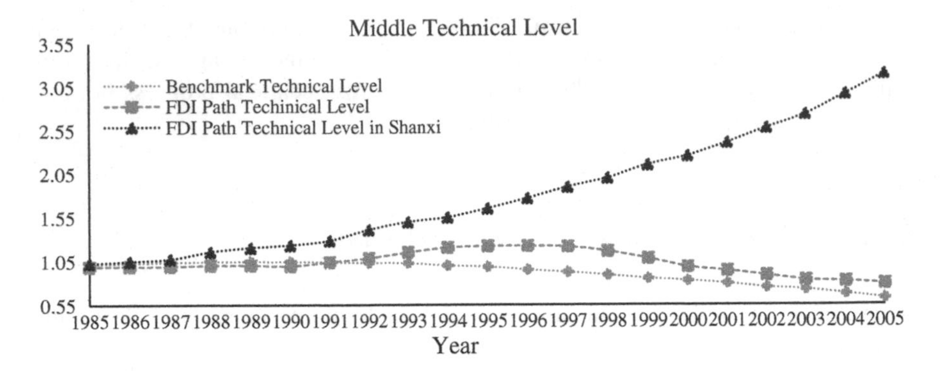

Fig. 5.14 Simulation of technical efficiency in Shanxi province

in Western region. The realization of increasing technical level may be obtained at least in Western provinces with their GDP per capita more than 0.6500 times of national level.

The GDP per capita in Shanxi (陕西) and Guangxi Provinces are 0.6400 and 0.6950 times of national GDP per capita in 1994. Implement the strategy with path-converged design, the technical level in Shanxi province witnesses a more violent decline and the technical level in Guangxi province successfully realizes steady increase. The facts demonstrate the 0.6500 times of national GDP per capita is a critical GDP per capita for Western region (Figs. 5.15, 5.16).

The realization of technical progress in Middle and Western regions validates the path-converged design can not only identify technical progress, but also can be applied to realize technical efficiency with perspective of management engineering.

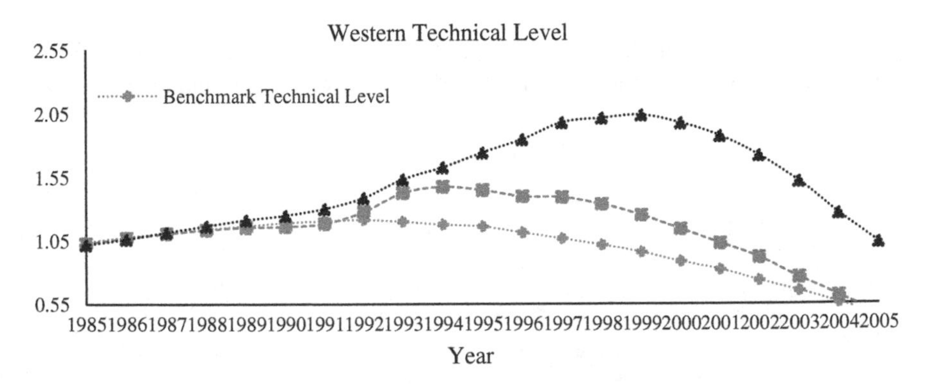

Fig. 5.15 Simulation of technical efficiency in Shanxi (陕西) province

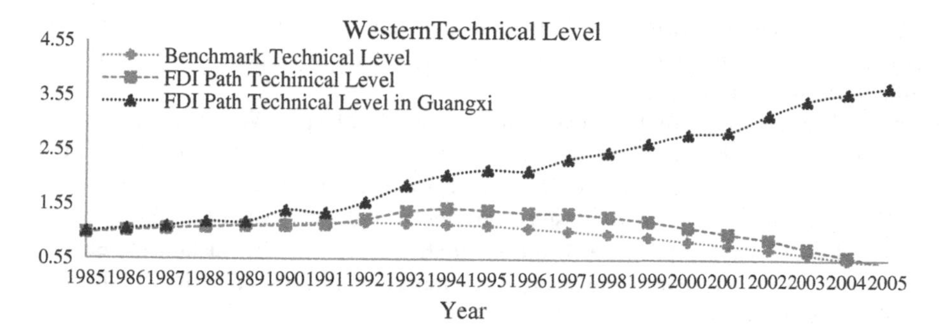

Fig. 5.16 Simulation of technical efficiency in Guangxi province

5.5 Conclusion

Most of previous researches have been presented to compute and analyze technical efficiency from economics and statistics perspectives. However, none of any application-based research to realize technical efficiency from management engineering perspective was found in literatures. This research establishes an innovative FDI Path-converged design technique not only to identify but also to realize change of technical efficiency in different regions of China.

The research in this chapter differs from previous researches with three aspects. First, it presents a benchmark model instead of production frontier surface in DEA and SFA as the criterion for measuring technical efficiency in regions. Second, the path-converged design technique not only identifies the change of technical efficiency by relative change of path-converged function based on benchmark function when FDI is added into original production function, but also identifies the changing trend of technical level and efficiency level over time. Most importantly, this research innovatively simulates change of technical efficiency in view of management engineering with path-converged design in both Middle and Western regions besides the identification of technical efficiency.

The empirical study with FDI path-***converged design*** technique among regions in China comes to the following three conclusions.

First, on the one hand, the technical progress with FDI path is significant in Eastern region after 1994 because its technical level increases successively after 1994, which reverses the declining trend during 1985–1994. Unfortunately, the same technical progress is not observed in Middle and Western regions since both their technical levels with FDI path and benchmark model are declining. On the other hand, the efficiency improvement with FDI path is slight in each region with slight positive differences are observed between the efficiency levels with FDI path and benchmark models in Eastern and Middle regions.

Second, the productivity growth with FDI path mainly attributes to technical progress rather than efficiency improvement in both Eastern and Middle regions since the difference of technical changes tends to greater than the difference of efficiency changes between the FDI path and benchmark model in each region. However, neither technical progress nor efficiency improvement is observed significantly in Western region.

Third, realizations of technical progress in Middle and Western regions are obtained with path-converged design. By implementing the strategy with path-converged design to middle provinces with more than 0.8000 times of national GDP per capita in 1996 and Western provinces with more than 0.6500 times of national GDP per capita in 1994, the simulations successfully sustain increasing technical levels and avoid the decreasing marginal returns in both Middle and Western regions.

References

Battese, G. E., & Coelli, T. J. (1995). A model for technical inefficiency effects in a stochastic frontier production function for panel data. *Empirical Economics, 20*(2), 325–332.

Bauer, P. (1990). Recent developments in the econometric estimation of frontiers. *Journal of Econometrics, 46*(1-2), 39–56.

Bitzer, J., & Kerekes, M. (2008). Does foreign direct investment transfer technology across borders? *New evidence, Economics Letters, 83*(3), 355–358.

Charnes, A., Cooper, W. W., & Rhodes, E. (1979). Measuring the efficiency of decision making units. *European Journal of Operational Research, 3*(4), 338–339.

Cook, W. D., & Seiford, L. M. (2009). Data envelopment analysis (DEA)-Thirty years on. *European Journal of Operational Research, 192*(1), 1–17.

Cook, P., & Uchida, Y. (2002). Productivity growth in east asia: a reappraisal. *Applied Economics, 34*, 1195–1207.

Cornwell, C., & Schmidt, P. (2008). Stochastic frontier analysis and efficiency estimation. In: *The Econometrics of Panel Data* (pp. 697–726), Part III, Vol. 46, Springer Berlin Heidelberg.

Fare, R., Grosskopf, S., & Lee, W. F. (2001). Productivity and technical change: The case of Taiwan. *Applied Economics, 33*(15), 1911–1924.

Farrell, M. J. (1957). The measurement of productive efficiency. *Journal of the Royal Statistical Society, Series A, 120*, 253–281.

Førsund, F., Lovell, C. A. K., & Schmidt, P. (1980). A survey of frontier production function and of their relationship to efficiency measurement. *Journal of Econometrics, 13*, 5–24.

Greene, W. H. (1993). The econometric approach to efficiency analysis. In H. Fried, C. A. K. Lovell, & S. Schmidt (Eds.), *The measurement of productive efficiency*. New York: Oxford University Press.

Kim, S., & Han, G. (2001). A decomposition of total factor productivity growth in Korean manufacturing industries: A stochastic frontier approach. *Journal of Productivity Analysis, 16*, 269–281.

Kim, T., & Park, C. (2006). Productivity growth in Korea: efficiency improvement or technical progress? *Applied Economics, 38*(8), 943–954.

Kumbhakar, S., & Lovell, C. A. K. (2000). *Stochastic frontier analysis*. Cambridge: Cambridge University Press.

Kuo, C–. C., & Yang, C.-H. (2008). Knowledge capital and spillover on regional economic growth: Evidence from China. *China Economic Review, 19*(4), 594–604.

Lai, M., Peng, S., & Bao, Q. (2006). Technology spillovers, absorptive capacity and economic growth. China Economic Review, 17, 300–320.

Lam, P., & Shiu, A. (2008). Productivity analysis of the telecommunications sector in China. *Telecommunications Policy, 32*(8), 559–571.

Lee, W. (2008). Benchmarking the energy efficiency of government buildingswith data envelopment analysis. *Energy and Buildings, 40*, 891–894.

Liu, Z. (2008). Foreign direct investment and technology spillovers: Theory and evidence. *Journal of Development Economics, 85*, 176–1938.

Lovell, C. A., & Schmidt, P. (1988). A comparison of alternative approaches to the measurement of productive efficiency. In: A. Dogramaci, & R. Färe (eds.) *Applications of modern production theory: Efficiency and production*. Boston: Kluwer Academic Publishers.

Luo, C. (2007). FDI, domestic capital and economic growth: Evidence from panel data at China's provincial level. *Frontiers of Economics in China, 2*(1), 92–113.

Ma, Y. (2006). The effect of FDI on China's economic growth. *Statistical Research, 3*, 51–54.

Ma, J., & Guo, W. (2007). The impact of agglomeration economy on FDI absorption in China. *Contemporary Finance & Economics, 8*, 109–112.

Mastromarco, C., & Ghosh, S. (2009). Foreign capital, human capital, and efficiency: A stochastic frontier analysis for developing countries. *World Development, 37*(2), 489–502.

Meeusen, W., & van den Broeck, J. (1977). Efficiency estimation from Cobb-Douglas production functions with composed error. *International Economic Review, 18*, 435–444.

Murakami, Y. (2007). Technology spillover from foreign-owned firms in Japanese manufacturing industry. *Journal of Asian Economics, 18*, 284–293.

Schmidt, P. (1985). Frontier production functions. *Econometric Reviews, 4*, 289–328.

Chapter 6
Changes in Scale and Technical Efficiency with Regional Transplantation Strategy

It is known to all that the growth in China has experienced severe regional difference among regions. Taken the perspective of production efficiency, it is necessary to identify whether the differences of scale and technical efficiency exist among regions, and if exist, how to change the difference would be rather important for equal and sustainable growth.

6.1 Introduction

Production efficiency measures whether the economy is producing as much as possible without wasting precious resources. Various researches are presented to estimate the production efficiency in view of Economics and Statistics. However, none of any application-based researches are presented to realize the production efficiency in the view of Management Engineering. When a new factor inflows into a production system, the change of production efficiency can be decomposed into the changes of allocation efficiency, scale efficiency and technical efficiency.

Regional discrepancy in China has been the focus of both governmental decision-makers and academic scholars. Figure 6.1 illustrates the widening regional discrepancy with the real GDP curves for Eastern, Middle and Western region during 1985–2005 at 1985s price. Numerous researches (Cai et al. 2002; Fu 2004; Lu and Lo 2008, etc.) have investigated the regional discrepancy in various frameworks. Specifically, majority of existing literatures on regional discrepancy focus on which factor (FDI, education, etc.) explains the regional discrepancy significantly (Wei 2002; Wu 2002; Fleisher et al. 2007) and fails to present whether the difference of scale efficiency or the gap of technical efficiency among regions due to the new factor attributes to the widening discrepancy.

FDI is thought to exert great positive roles on growth in China (Lai et al. 2006; Luo 2007). Debates on whether the regional absorption differences of FDI significantly contribute to the regional discrepancy remain continuous and fail to come to definite conclusion. Figure 6.2 presents the FDI proportion to GDP for Eastern, Middle and Western regions during 1985–2004. The FDI proportion in

B. Xu et al., *Changes in Production Efficiency in China*,
DOI: 10.1007/978-1-4614-7720-4_6,

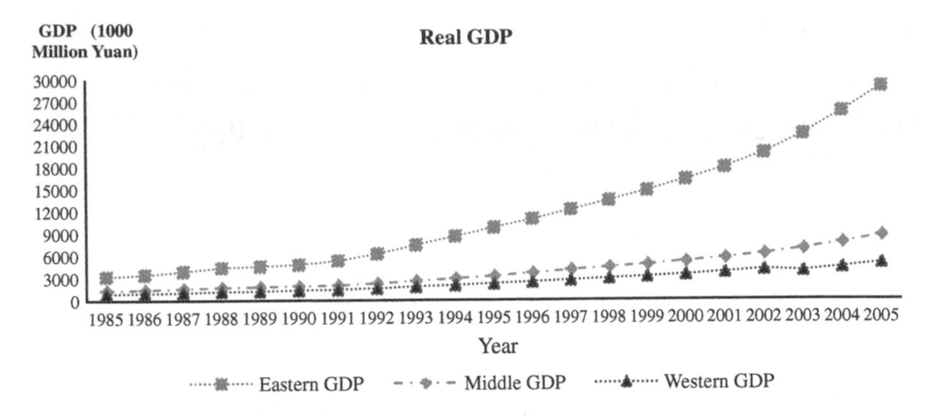

Fig. 6.1 Real GDP for eastern, middle and western region during 1985–2005

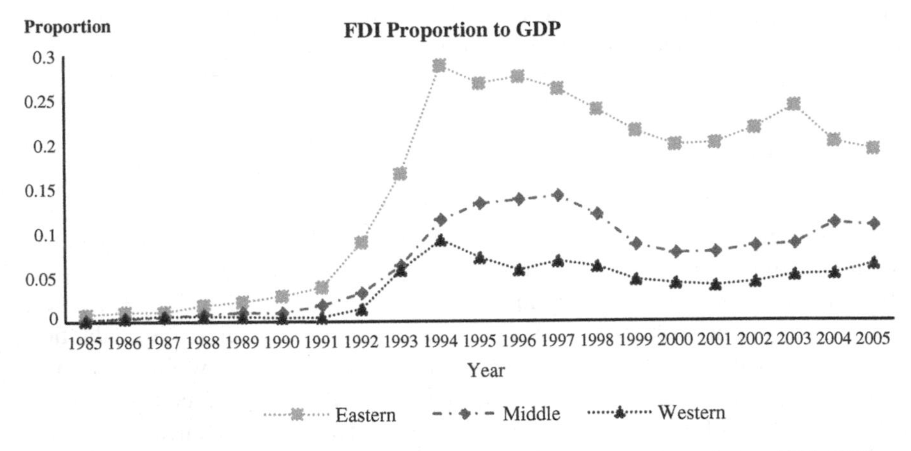

Fig. 6.2 FDI proportion to GDP for eastern, middle and western regions during 1985–2005

Eastern region is the highest and the one in Western region is the lowest. Despite of the regional discrepancy of the proportion, the amount of FDI is relative small compared with GDP in each region. In this sense, FDI exerts its positive role on growth through improving technical efficiency (Luo 2007) rather than bringing about scale efficiency.

Different from the previous literatures, this research firstly focuses on the identification of regional discrepancies of technical efficiency and scale efficiency, and secondly emphasizes on the realization of changes of production efficiency with changes of technical efficiency and scale efficiency through nonparametric path-converged approach. To reach the objectives of identification and realization, three steps are introduced in the research. First, the underlying structure of economic growth needs to be estimated. Second, the path needs to be selected to identify the technical efficiency and scale efficiency in each region, which requires the establishment of both path model and benchmark model of underlying production function and realizes the convergence. Third, the production efficiency needs to be realized with the changes of technical efficiency and scale efficiency.

On the one hand, a benchmark production function is presented to estimate the underlying structure with common statistic approach. On the other hand, path model of production function is also obtained with the selected path. Furthermore, the scale efficiency and technical efficiency due to the path can be investigated. The scale efficiencies emphasize on the output elasticities of inputs, while the technical efficiency focuses on total factor productivity (TFP) of the inputs.

Path selection is of crucial importance to implement projects. The path selection is characterized with production efficiency promotion. FDI is employed as the path in the research because empirical researches yield a robust and theoretically consistent results that FDI affects production efficiency with the channels of total factor productivity (Liu 2002; Javorcik 2004; Goss et al. 2007; Murakami 2004) and technological spillover (Lee 2006; Fan and Hu 2007).

With the issues on production function estimation and path selection solved, realizations on change of production efficiency can be presented. Based on time-varying production function with nonparametric FDI path-converged approach, this research first identifies the regional discrepancies of technical efficiency and scale efficiency in China, and second realizes the changing production efficiency by changes of technical efficiency and scale efficiency.

The rest of this chapter is organized as follows. Section 6.2 presents the nonparametric path-converged design with time-varying production function. Section 6.3 gives the path identification of regional technical efficiency and scale efficiency. Section 6.4 realizes the change of production efficiency with changes of technical efficiency and scale efficiency. And Sect. 6.5 finally concludes.

6.2 Path-Converged Design

Step A. **Benchmark model**

Assume the well-known C-D production function as the underlying structure S_0:

$$Y(t) = A_0(t)K(t)^{\alpha}L(t)^{\beta} \tag{6.1}$$

where K and L denotes the inputs of labor and capital, α and β correspond to output elasticities of capital and labor respectively, A_0 refers to the technical level of production.

Define a benchmark model S_1:

$$Y_1(t) = K(t)^{\alpha(t)}L(t)^{\beta(t)} \tag{6.2}$$

Step B. **Path model S_2:**

To depict the change process, define the path model S_2:

$$Y_2(t) = FDI(t)^{\lambda(t)} K(t)^{\alpha(t)} L(t)^{\beta(t)} \tag{6.3}$$

Rewriting Eq. (6.1), it is shown that

$$\begin{aligned} Y_1(t) &= K(t)^{\alpha(t)} L(t)^{\beta(t)} \\ &= K(t)^{\alpha_1 t - \alpha_1 t_0} L(t)^{\beta_1 t - \beta_1 t_0} (K(t)L(t))^{o(t)} K(t)^{\alpha_0} L(t)^{\beta_0} \\ &= A_1(t, t_0) K(t)^{\alpha_0} L(t)^{\beta_0} \end{aligned} \tag{6.4}$$

where α_0 and β_0 are taken as capital and labor elasticity respectively, $A_1(t, t_0)K(t)^{\alpha_1 t - \alpha_1 t_0} L(t)^{\beta_1 t - \beta_1 t_0} (K(t)L(t))^{o(t-t_0)}$ is the total factor productivity.

Rewriting Eq. (6.2), it is shown that

$$\begin{aligned} Y_2(t) &= FDI(t)^{\lambda(t)} K(t)^{\alpha(t)} L(t)^{\beta(t)} \\ &= FDI(t)^{\lambda_1 t - \lambda_1 t_0} K(t)^{\alpha_1 t - \alpha_1 t_0} L(t)^{\beta_1 t - \beta_1 t_0} (FDI(t)K(t)L(t))^{o(t-t_0)} FDI(t)^{\lambda_0} K(t)^{\alpha_0} L(t)^{\beta_0} \\ &= A_2(t, t_0) FDI(t)^{\lambda_0} K(t)^{\alpha_0} L(t)^{\beta_0} \end{aligned} \tag{6.5}$$

where λ_0 is output elasticity of FDI. $\alpha_0 + \beta_0 + \lambda_0$ represents the scale efficiency.

Models (6.4) and (6.5) can be estimated by the local weighed least-squares technique:

$$\min_{\theta(t_0)} \sum_{t=1}^{n} (\ln Y_t - (\lambda_0 + \lambda_1 (t - t_0)) FDI_t - (\alpha_0 + \alpha_1 (t - t_0)) \ln K_t - (\beta_0 + \beta_1 (t - t_0)) \ln L_t)^2 K_h(t - t_0)$$

where $\theta(t_0) = (\lambda_0, \alpha_0, \beta_0, \lambda_1, \alpha_1, \beta_1)^T$, $K_h(x) = h^{-1}K(x/h)$ and h are kernel functions and bandwidth.

The bandwidth in this research is selected by the least squares cross-validation method (Stone 1984), and kernel function is given by Gauss kernel function:

$$K(x) = (2\pi)^{-1/2} \exp(-x^2/2)$$

The total factor productivity here is given by

$$A_2(t, t_0) = FDI(t)^{\lambda_1 t - \lambda_1 t_0} K(t)^{\alpha_1 t - \alpha_1 t_0} L(t)^{\beta_1 t - \beta_1 t_0} (FDI(t)K(t)L(t))^{o(t-t_0)}$$

The technical change and efficiency level of TFP of Model (6.2) are estimated as follows.

The technical change at time t based on time t_0 is:

$$\left(FDI(t)^{\lambda_1 t - \lambda_1 t_0} K(t)^{\alpha_1 t - \alpha_1 t_0} L(t)^{\beta_1 t - \beta_1 t_0} FDI(t_0)^{\lambda_1 t - \lambda_1 t_0} K(t_0)^{\alpha_1 t - \alpha_1 t_0} L(t_0)^{\beta_1 t - \beta_1 t_0} \right)^{1/2}$$

And the technical level is given by:

$$\hat{T}ech(t) = FDI^{\hat{\lambda}_1 t}(t)K^{\hat{\alpha}_1 t}(t)L^{\hat{\beta}_1 t}(t) \tag{6.6}$$

The efficiency level is given by:

$$E\ddot{o}ff(t) = Y(t)\left(FDI(t)^{\lambda\ddot{o}_0}K(t)^{\alpha\ddot{o}_0}L(t)^{\beta\ddot{o}_0}\right)^{-1} \tag{6.7}$$

The efficiency change at point t to t_0 is obtained by the rate of estimator (6.7).

If the efficiency level equals to 1, then the total factor productivity turns out to be the technical change, i.e., technical inefficiency does not exist. The above decomposition implies that technical inefficiency does not affect total factor productivity growth, as in the **Solow residual approach**.

6.3 Identification of Regional Technical and Scale Efficiency

The data used in this chapter is the same with the one in Chap. 4. The path identification of technical efficiency and scale efficiency are presented in Figs. 6.3 and 6.4. The introduction of FDI path into the economic system causes significant regional discrepancy in both technical efficiency and scale efficiency. Since the efficiency levels almost equal to 1 for all regions, the technical level in Fig. 6.3 is approximately taken as the technical efficiency in this research. The scale effect in Fig. 6.4 is taken as scale efficiency with the FDI path.

Figure 6.3 lists the technical levels with FDI path among regions, which are estimated by model (6.6). The study concerns the varying trend of technical level rather than the absolute level of technical level, because the increasing trend of technical level will successfully avoids the diminishing marginal returns of capital and labor inputs. With the FDI path included, the technical levels among regions

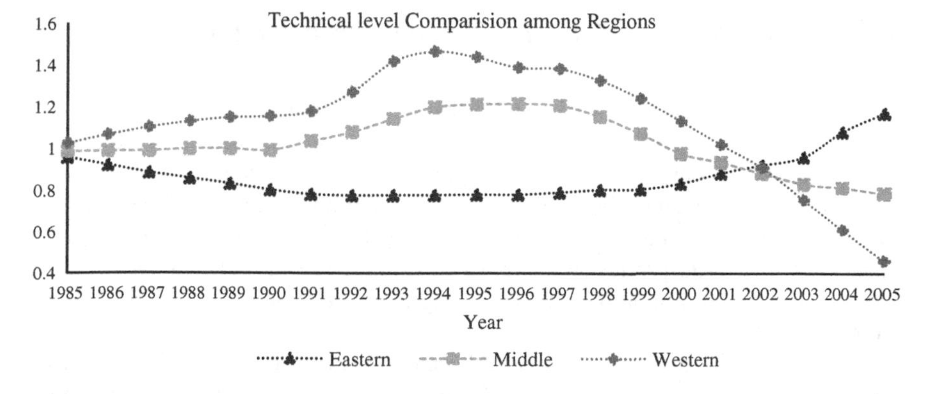

Fig. 6.3 Technical levels comparison among regions

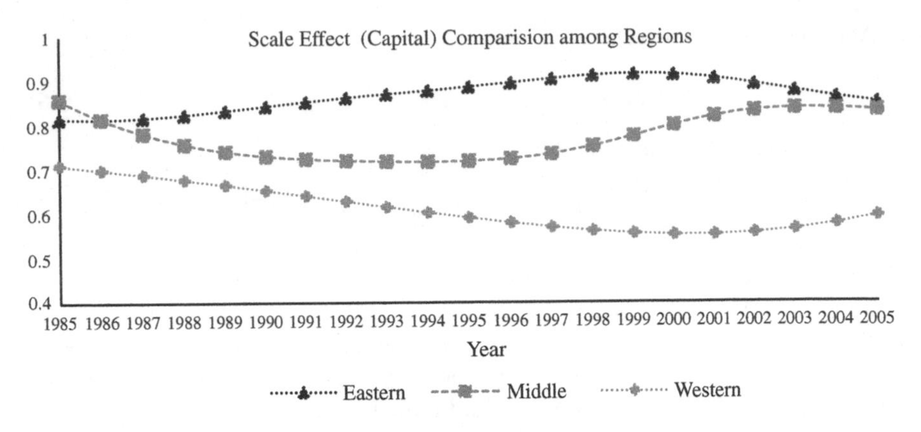

Fig. 6.4 Scale effects (capital) comparison among regions

witness significant regional discrepancy. The technical level in Eastern region witnesses an increasing trend and the ones in both Middle and Western regions experience a declining trend.

Figure 6.4 illustrates the scale effects with FDI path among regions, which are the estimations of capital elasticity. Because the growth in China is mostly investment-oriented; the capital elasticity explains the major part of scale effect. Figure 6.4 shows the scale effect in Eastern region is the greatest with an increasing trend and the one in western region is the lowest with a declining trend.

The path identification illustrates the regional discrepancy in both technical efficiency and scale efficiency. Since both technical promotion and increasing returns to scale are the pursuits for each region, the identification implies possible realization on production efficiency.

6.4 Transplantation Strategy of Efficiency Change

6.4.1 Strategy with Technical Efficiency Transplantation

The path identification with FDI path illustrates that the technical efficiency in Eastern region witnesses an increasing trend whiles the ones in both Middle and Western regions experience a declining trend. To keep the technical level from declining is the first step for both regions to achieve output growth in their learning process, which is the key to prevent the inputs free from decreasing marginal returns and to pursue productivity improvements. The improvement of productivity is of crucial importance to the long-term development and prosperity (Hannula 2002).

The realized process is to transplant the technical efficiency in Eastern region into both Middle and Western regions. To measure how much output growth will

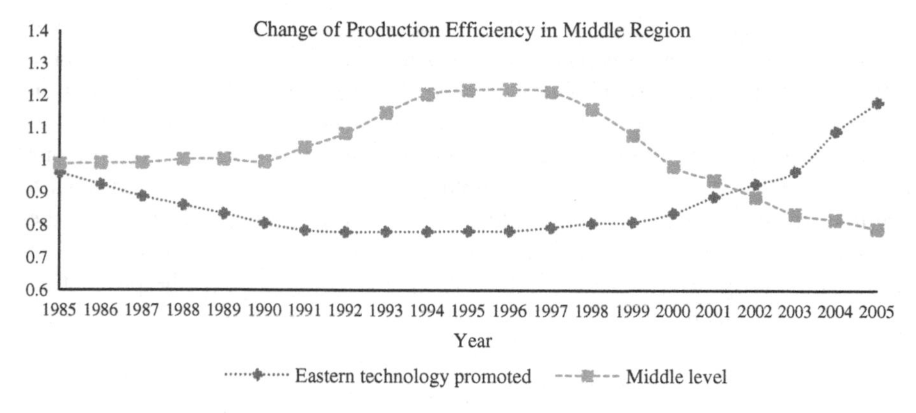

Fig. 6.5 Change of production efficiency with technical transplantation in middle region

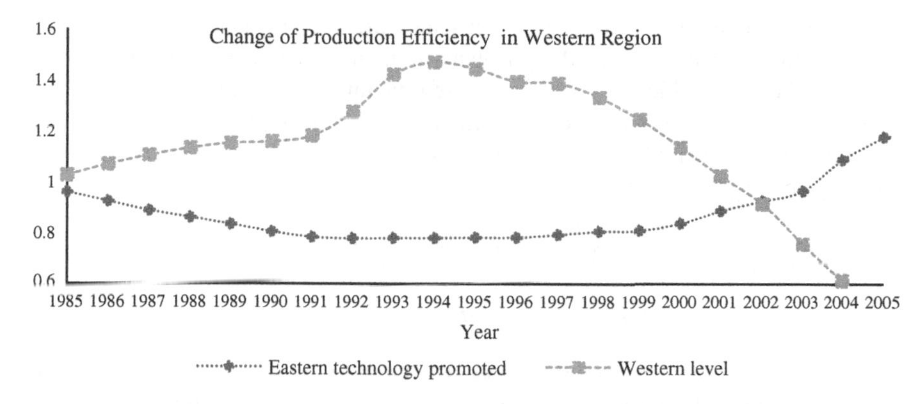

Fig. 6.6 Change of production efficiency with technical transplantation in western region

occur with the transplantation, the difference between Models (6.8) and (6.9) gives the changes of production efficiency in Middle (Western regions).

$$A_E(K(t)L(t))^{o(t-t_0)_1}FDI^{\lambda_0}(t)K^{\alpha_0}(t)L^{\beta_0}(t)Y^{-1}(t) \tag{6.8}$$

$$A_0(K(t)L(t))^{o(t-t_0)_1}FDI^{\lambda_0}(t)K^{\alpha_0}(t)L^{\beta_0}(t)Y^{-1}(t) \tag{6.9}$$

Except for replacing the Middle (Western) technical level A_0 by the Eastern technical level A_E, both the inputs and the output elasticity of capital, labor and FDI path keep the same.

Figures 6.5 and 6.6 present the changes of production efficiency in Middle and Western regions if the Eastern technological level is transplanted into both regions, i.e., when the realization succeed in Middle and Western regions.

The change of production efficiency is perceived by the differences between the two curves in each figure. By studying the developed technology in Eastern region, both Middle and Western region will go through a varying process in production efficiency. The absorption capability of new technology during the former stage (1985–2002) is rather small and even negative because of the investment in education, human capital, and so on. Fortunately, the learning process starts to present positive spillover effect during the latter stage (2003–2005) in both regions because the curves with Eastern technology promoted lie upon the ones with original technical level in both Middle and Western regions, illuminating the transplantation of Eastern technical level into Middle and Western regions not only is promising in bringing about output growth, but also keeps the production efficiency from decreasing.

6.4.2 Strategy with Scale Efficiency Transplantation

The Path identification illustrates that the scale efficiency in Eastern region is the highest. Similar to the technical transplantation, the realized process here is to transplant the scale efficiency in Eastern region into both Middle and Western regions. To measure how much output growth will occur with the transplantation, the difference between Models (6.10) and (6.11) gives the changes of production efficiency in Middle (Western regions).

$$(TFP_0)\left(K^{E}_{\alpha_0}(t)\right)FDI^{\lambda_0^E}(t)L^{\beta_0^E}(t)Y^{-1}(t) \tag{6.10}$$

$$(TFP_0)(K^{\alpha_0}(t))FDI^{\lambda_0}(t)L^{\beta_0}(t)Y^{-1}(t) \tag{6.11}$$

Except for replacing the Middle (Western) output elasticities α_0, β_0 and λ_0 by the Eastern output elasticities α_0^E, β_0^E and λ_0^E respectively, both the inputs and total factor productivity keep the same.

Figures 6.7 and 6.8 present the changes of production efficiency in Middle and Western regions if the Eastern scale efficiency is transplanted into both regions. By observing the differences between the two curves in each figure, it can be concluded that the production efficiency in both regions are promoted by transplanting the eastern scale effect into Middle and Western regions. However, the original declining trends of production efficiency in both regions remain unchanged.

6.4.3 Strategy with Joint Transplantation of Technical and Scale Efficiencies

The realized process here is to transplant both the technical efficiency and scale efficiency in Eastern region into both Middle and Western regions. To measure how much output growth will occur with the transplantation, the difference

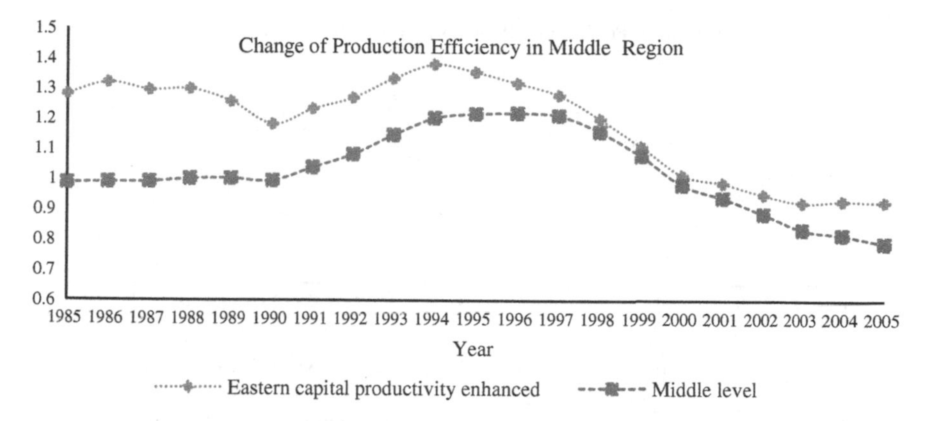

Fig. 6.7 Change of production efficiency with scale efficiency transplantation in middle region

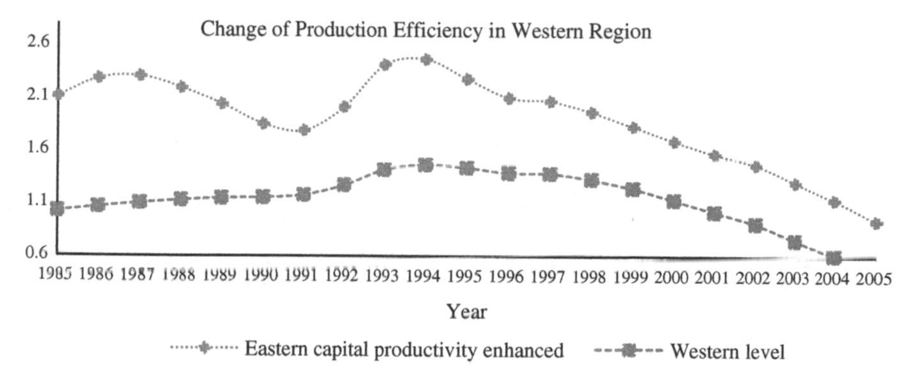

Fig. 6.8 Change of production efficiency with scale efficiency transplantation in western region

between Models (6.12) and (6.13) gives the changes of production efficiency in Middle (Western regions).

$$A_E K_{\alpha_0}^{E} (K(t)L(t))^{o(t-t_0)_E} FDI^{\lambda_E}(t)(t)L^{\beta_E}(t)Y^{-1}(t) \tag{6.12}$$

$$A_0 K_{\alpha_0}^{1} (K(t)L(t))^{o(t-t_0)_1} FDI^{\lambda_0}(t)(t)L^{\beta_0}(t)Y^{-1}(t) \tag{6.13}$$

Except for the input factors of the capital, labor and FDI path in Middle and Western region keeping unchanged, both the technical efficiency and output elasticities in Middle (Western) region are replaced by the ones in Eastern region respectively.

Figures 6.9 and 6.10 illustrate the changes of production efficiency in Middle and Western regions if both the scale efficiency and technical efficiency in Eastern region are transplanted into Middle and Western regions. Both figures show that the production efficiency is promoted more greatly when both transplantations are implemented into Middle and Western regions. The output growths in both regions

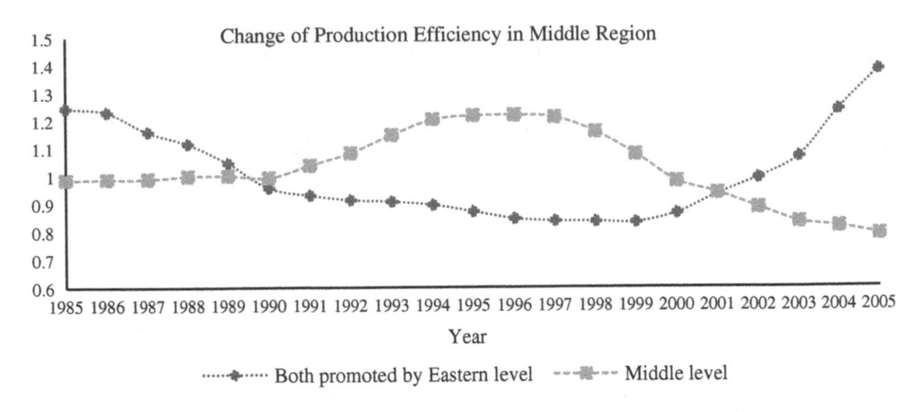

Fig. 6.9 Change of production efficiency with both transplantations in middle region

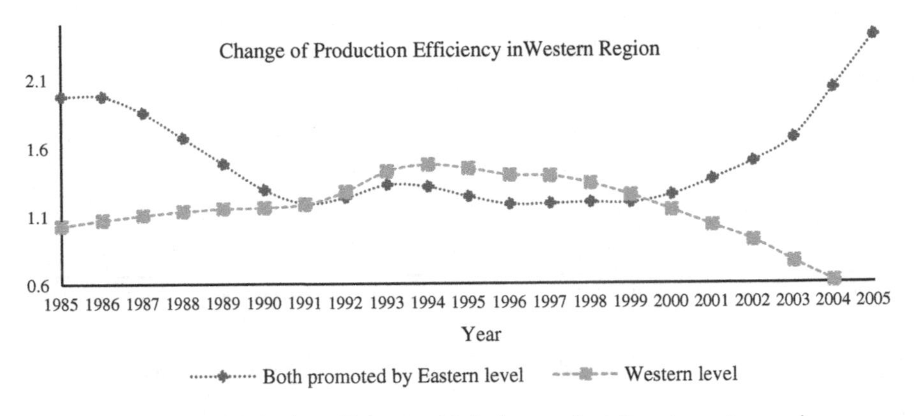

Fig. 6.10 Change of production efficiency with both transplantations in western region

with both impacts are the synthesized output growths with the respective impacts of scale efficiency and technical level. The fact implies that the idea of learning technical level and pursuing scale effects in Eastern region for Middle and Western regions performs well in bringing about output growth in both regions, which provides sound ground for higher authorities to make decisions.

6.4.4 Feasibility Analysis of Transplantation

It is natural to raise the problem whether the transplantation are feasible. First, the transplantations with scale efficiency are feasible since the study in Sect. 4.5 validates both Middle and Western region can realize the crowd-in effect of FDI path in provinces with their urbanization levels higher than 0.2500. Second, the transplantations with technical efficiency are feasible since the study in Sect. 5.4

presents that technical progress can be realized in Middle provinces with more than 0.8000 times of national GDP per capita in 1996 and Western provinces with more than 0.6500 times of national GDP per capita in 1994.

6.5 Conclusion

The study innovatively presents several realization processes to realize the production efficiency in view of Management Engineering. Based on time-varying production function with FDI path-converged approach, this research firstly identify the regional discrepancies of technical efficiency and scale efficiency, and secondly realizes the changes of production efficiency with the changes of technical efficiency and scale efficiency in Middle and Western regions.

The identification and realization come to three main conclusions.

First, the path identification illustrates the regional discrepancy in both technical efficiency and scale efficiency. On the one hand, the technical efficiency in Eastern region is increasing while the ones in Middle and Western regions are declining. On the other hand, the scale efficiency in Eastern region is the highest among regions. Since both technical promotion and increasing returns to scale are the pursuits for each region, the identification implies possible realization on production efficiency.

Second, the transplantation of Eastern technical level into Middle and Western regions not only is promising in bringing about output growth, but also keeps the production efficiency from decreasing. The transplantation of the scale efficiency in Eastern region into Middle and Western regions succeeds in promoting production efficiencies in both regions, while fails in reversing the original declining trends of production efficiency in both regions. The transplantation of both technical efficiency and scale efficiency in Eastern region into Middle and Western regions illustrate synthesized effect of respective transplantation.

Third, the simulations are feasible with the environment of urbanization development and the growth of GDP per capita level. However, how to realize the realizations with the design of urbanization environment and growth of GDP per capita needs to be further explored.

References

Cai, F., Wang, D., & Du, Y. (2002). Regional disparity and economic growth in China, the impact of labor market distortions. *China Economic Review, 13*, 197–212.

Fan, C., & Hu, Y. (2007). Foreign direct investment and indigenous technological efforts: Evidence from China. *Economics Letters, 96*, 253–258.

Fleisher B, Li H, Zhao M (2007) Human capital, economic growth, and regional inequality in China. William Davidson Institute Working Papers Series, No wp857.

Fu, X. (2004). Limited linkages from growth engines and regional disparities in China. *Journal of Comparative Economics, 32*, 148–164.

Goss, E., Wingender, J. R., & Torau, J. M. (2007). The contribution of foreign capital to U.S. productivity growth. *The Quarterly Review of Economics and Finance, 47*(3), 383–396.

Hannula, M. (2002). Total productivity measurement based on partial productivity ratios. *International Journal of Production Economics, 78*(1), 57–67.

Javorcik, B. (2004). Does foreign direct investment increase the productivity of domestic firms? In search of spillovers through backward linkages. *American Economic Review, 94*, 605–627.

Lai, M., Peng, S., & Bao, Q. (2006). Technology spillovers, absorptive capacity and economic growth. *China Economic Review, 17*, 300–320.

Luo, C. (2007). FDI, domestic capital and economic growth: Evidence from panel data at China's provincial level. *Frontiers of Economics in China, 2*(1), 92–113.

Liu, Z. (2002). Foreign direct investment and technology spillover: Evidence from China. *Journal of Comparative Economics, 30*, 579–602.

Lee, G. (2006). The effectiveness of international knowledge spillover channels. *European Economic Review, 50*, 2075–2088.

Stone, C. J. (1984). An asymptotically optimal window selection rule for kernel density estimates. *The Annals of Statistics, 12*, 1285–1297.

Wei, H. K. (2002). Effects of foreign direct investment on regional economic growth in China. *Journal of Economic Research, 4*, 19–26.

Wu, J. (2002). Regional discrepancy of FDI in China and its effect on economic growth. *Journal of Economic Research, 4*, 27–34.

Chapter 7
Sustainable Efficiency in China with Urbanization

Under the background of sustainable growth, Chinese government has put forward both stimulating domestic private investment and accelerating healthy urbanization development are as priority tasks in following five years in 2013. Since urbanization is taken as an effective and necessary path to pursue growth, Chap. 3 measures the allocation efficiency under the background of urbanization with proposed strategies of optimal population migration in urban planning in China. Investment is taken as main factor in driving growth, Chaps. 2, 4 and 5 measures the changes in allocation efficiency, scale efficiency, and technical efficiency with designed investment strategies to explore how investment changes production efficiency. Taken the perspective of regional equity, Chap. 6 designs regional transplantation strategies of investment to measure the changes in both scale and technical efficiency for growth in order to conquer severe regional difference in China.

Similar to the regional difference in scale and technical efficiency due to investment, the urbanization gaps among different regions and different city groups (big, medium, and small cities) will inevitably bring about imbalance in production efficiency. Synthesizing the regional differences in investment and urbanization developments, the transplantation strategies with both investment and population agglomeration between different regions and city groups will be more comprehensive in measuring the changes in production efficiency under sustainable growth background. Therefore, this chapter aims to pursue sustainable production efficiency in China with optimal resource allocation of investment and balanced urbanization development.

7.1 Introduction

Chen (2006) and Li (2007) assert a strong correlation between urbanization and income disparities across China's 31 provinces, autonomous regions, and province-level municipalities regions, as well as between urban and rural residents within each region. Provinces with a higher level of urbanization also have higher

B. Xu et al., *Changes in Production Efficiency in China*,

DOI: 10.1007/978-1-4614-7720-4_7,

per capita GDP and less urban–rural income disparity. Since the modern cities are the core of regional economy, Lu and Chen (2004) hold the urbanization significantly decreases the statistical urban–rural inequality.

Liu et al. (2006) testify the great role of the investment in fixed assets in promoting urbanization. Li et al. (2005) give the quantitative analysis of the infrastructure of investment in the process of urbanization. Catin et al. (2005) argue that the government should appropriately control less to internal migration reducing the regional disparities. Chen (2006) emphasizes the role of government in urbanizing. Zhu (2006) proposes the most effective measure to reduce the disequilibrium of regional development in China is to accelerate the urbanization process in middle and west regions.

Existing literatures pay little attentions on providing practical strategies to promote urbanization level besides several qualitative suggestions. Reducing the disequilibrium of regional development and realizing the optimization of resource allocation will be severe challenging alternatives (Zheng et al. 2007) for pursuing harmonious and sustainable economic growth in China. It is testified that urbanization is a vigorous drive of Chinese economic growth and produces the most important economic benefits from the endogenous, factor, path and exogenous perspectives. And population migration is taken as a key factor that works on the urbanization development.

This chapter takes urbanization as the necessary path to pursue sustainable production efficiency in china. It designs several transplantation strategies with both investment and population agglomeration to reduce urbanization gap among regions and realize optimization of resource allocation among city groups. With the optimal resource allocation and balanced urbanization development, sustainable production efficiency will be realized. As a unitary country, the Chinese government is probably relatively unconstrained to pursue appropriate strategies and instruments to reach sustainable growth via the convergence in regional urbanization development and realize the optimization of resource allocation for promoting urbanization in different city groups.

The rest of the chapter is outlined as follows. Section 7.2 presents the design of urbanization path. Section 7.3 designs several strategies for balanced development among eastern, middle and western regions. Section 7.4 provides the strategies to realize the optimization of resource allocation for promoting urbanization among small, medium-sized and big cities. Section 7.5 presents the feasibility analysis on the strategies and Sect. 7.6 briefly concludes.

7.2 Design of Urbanization Path

Some strategies of urbanization development are designed by factor path driven. Underlying path structure S_0, benchmark path structure S_1 and factor path structure S_2 are given as:

$$\text{Underlying path } S_0 : f(x)$$
$$\text{Benchmark path } S_1 : f_n(x) = \frac{1}{n}\sum_{i=1}^{n} h^{-1}K\left(\frac{x-X_i}{h}\right)$$
$$\text{Factor path } S_2 : f_{\omega n}(x) = \sum_{j=1}^{n} \omega_j h^{-1}K\left(\frac{x-X_i}{h}\right) \tag{7.1}$$

where $K(\cdot)$ is a Gaussian kernel function and h is selected according to Cross-Validation method. ω_i denotes the population (investment) weight of city i, $i = 1, \cdots, n$. The benchmark path structure for urbanization development is presented by means of the kernel density estimation. And the factor path structure is one by means of the weighted kernel density estimation.

Referring to the studies of Kim and Lee (2005) or Xu and Cai (2007), given some mild regular conditions, it is easy to see that:

$$P\left(\lim_{n\to\infty}[f_n(x) - f(x)] = 0\right) = 1, \text{and}$$
$$P\left(\lim_{n\to\infty}[f_{\omega n}(x) - f(x)] = 0\right) = 1$$

Result in
$$P\left(\lim_{n\to\infty}[f_{\omega n}(x) - f_n(x)] = 0\right) = 1.$$

In other words, the underlying path structure S_0 is *identifiable* using factor ω, that is, the underlying path structure S_0 is identified by the difference between benchmark path S_1 and factor path S_2.

7.3 Balanced Development Among Regions

The data employed to identify urbanization developments in different regions is obtained from China City Statistical Yearbook 2005. It provides the cross section data of 282 prefecture-level cities in 2004. The indexes are citywide total population at the end of 2004, downtown total population at the end of 2004, citywide nonagricultural population, downtown nonagricultural population of 2004, downtown gross fixed-assets investment of 2004, downtown total output value of 2004 (Unit: 10,000 Yuan) in each prefecture-level city.

According to the classification criterion of China City Statistical Yearbook 2005, the *Eastern region* contains Beijing, Tianjin, Hebei, Liaoning, Shanghai, Jiangsu, Zhejiang, Fujian, Shandong, Guangdong and Hainan province; the *Middle region* contains Shanxi, Neimenggu, Jilin, Heilongjiang, Anhui, Jiangxi, Henan, Hubei, Hunan; and the *Western region* contains Guangxi, Chongqing, Sichuan, Guizhou, Yunnan, Xizang, Shanxi, Gansu, Ningxia, Qinghai and Xinjiang. Thus the 282 prefecture-level cities can be classified into three groups: eastern cities, middle cities and western cities. The urbanization of an individual prefecture-level city is defined as the proportion of its citywide nonagricultural population for 2004 to its citywide total population at the end of 2004.

The second data to identify urbanization developments for different city sizes come from the China City Statistical Yearbooks 2004 and 2005. The Yearbook 2004 provides the cross section data for 276 prefecture-level cities in 2003. To validate the role of investment on urbanization development, the identification in different city groups keeps the 276 cities the same for 2004. Besides the same indexes for 2003 and 2004 as the first data source, the downtown gross fixed-assets investments in 2003 and 2004 for each prefecture-level city are included. According to the classification criterion of China City Statistical Yearbook 2004, the *small city* is defined as its downtown total population less than 500,000, the *medium-sized city* 500,000 to 1 million, and the *big city* larger than 1 million. The cities at a prefecture level in 2003 include 102 big cities, 108 medium-sized cities and 66 small cities. The study holds each city group in 2004 the same in spite of the population change.

Considering the population migration and investment both changes production efficiency, the strategies designed for reducing the urbanization gap in this chapter are concerning population migration and strategies concerning investment enhancement.

p_{0i}, I_{0i} and g_i are the original total downtown population, original downtown investment and total output value (GDP) of middle city i. The average urbanization development levels of population factor path and investment factor path in middle region are 0.351 and 0.392, respectively.

The strategies designed in this chapter rely on the change of population and investment in fixed assets. On one hand, several strategies of population migration and investment enhancement are designed to reduce the urbanization gap among regions. On the other hand, strategies of investment are designed to optimize resource allocation for urbanization promotion among small, medium-sized and big cities. Both cases will pursue sustainable growth through the urbanization promotion.

7.3.1 Strategies of Population Migration

Strategy 1. Exterior Immigration with Average Increase

Increase the population of a middle region by 20 (or 30) percent of its original total population. The immigration to each city in middle region is distributed by the ratio of its GDP to the total GDP, i.e.

$$p_{1i} = p_{0i} + 0.2 \times \sum_{j=1}^{n} p_{0j} \times \frac{g_i}{\sum_{j=1}^{n} g_j}, \quad p_{2i} = p_{0i} + 0.3 \times \sum_{j=1}^{n} p_{0j} \times \frac{g_i}{\sum_{j=1}^{n} g_j} \tag{7.2}$$

where p_{1i} and p_{2i} are the respective population in middle city i when the population increased by 20 and 30 %. $\omega_{1i} = p_{1i}/\sum_{j=1}^{n} p_{1j}$ and $\omega_{2i} = p_{2i}/\sum_{j=1}^{n} p_{2j}$ are the new weights for middle city i when the population increased by 20 (or 30) percent with project 1.

Figure 7.1 illustrates the comparison of new population (increased by 30 %) factor path in a middle region, original population factor paths in middle and eastern region. It shows a slight promotion in the urbanization of the middle region. Table 7.1 presents the effect of Strategy 1.

$f_{\omega 0,h}(x)$ denotes the original population factor path in Middle region, $f_{\omega 1,h}(x)$ the increased population (by 20 %) factor path in Middle region, $f_{\omega 2,h}(x)$ the increased population (by 30 %) factor path in Middle region, $f'_{\omega 0,h}(x)$ the original population factor path in eastern region, $S_1 = \int_0^{0.35} f(x)dx$, $S_5 = \int_{0.35}^{1} f(x)dx$ and $EX = \int_0^{\infty} xf_h(x)dx$.

Considering the change of the average urbanization level in Middle region and the integral value change of S_1 and S_5, Strategy 1 shows a slight promotion of urbanization in middle region and has a little effect on reducing the urbanization gap between the middle and eastern regions.

Strategy 2. Exterior Immigration with Selected Increase

Increase the population of middle cities whose urbanization level is higher than the average urbanization level through the original population factor path by 20 (or 30) percent of its original total population, respectively. The immigration to each city in Middle region is distributed by the ratio of its GDP to the total GDP of middle cities whose urbanization levels are higher than the average level, i.e.

$$p_{1i} = p_{0i} + \frac{g_i I(L_i \geq a)}{\sum_{j=1}^{n} g_j I(L_j \geq a)} \times 0.2 \sum_{k=1}^{n} p_{0k}, \quad p_{2i} = p_{0i} + \frac{g_i I(L_i \geq a)}{\sum_{j=1}^{n} g_j I(L_j \geq a)} \times 0.3 \sum_{k=1}^{n} p_{0k} \tag{7.3}$$

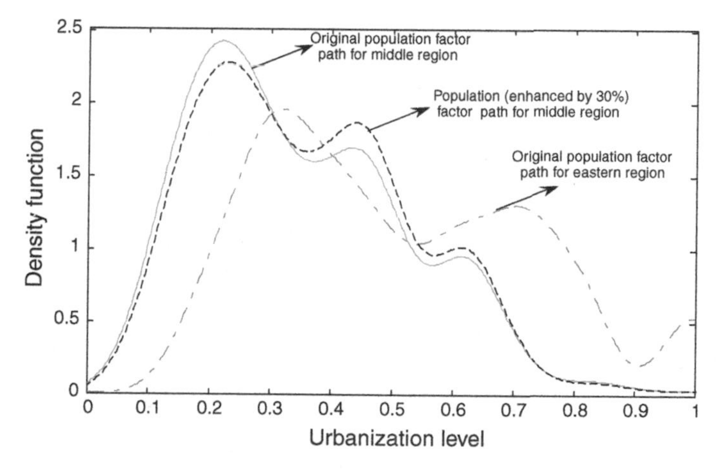

Fig. 7.1 Population factor paths in middle region with Strategy 1

Table 7.1 Effect of Strategy 1

$f(x)$	S_1	S_5	EX
$f_{\omega 0,h}(x)$	0.546	0.454	0.351
$f_{\omega 1,h}(x)$	0.521	0.479	0.359
$f_{\omega 2,h}(x)$	0.512	0.488	0.362
$f'_{\omega 0,h}(x)$	0.331	0.669	0.483

where $I(\cdot) = 1$ or 0 if $L_j \geq a$ or other, stands for the indicator function. L_j be the urbanization levels in city j, $a = EX = 0.3514$ $\omega_{1i} = p_{1i}/\sum_{j=1}^{n} p_{1j}$ and $\omega_{2i} = p_{2i}/\sum_{j=1}^{n} p_{2j}$ are the new weights for middle city i where the population increased by 20 and 30 % with project 2.

Figure 7.2 displays the comparison of new population (increased by 30 % with Strategy 2) factor path for middle region, original population factor paths for middle and eastern region. Table 7.2 gives the effect of Strategy 2.

$f_{\omega 1,h}(x)$ is the increased population (by 20 % with Strategy 2) factor path in middle region, $f_{\omega 2,h}(x)$ the increased population (by 30 % with Strategy 2) factor path in middle region. $f_{\omega 0,h}(x)$, $f'_{\omega 0,h}(x)$, S_1, S_5 and EX are the same as the indexes in Table 7.1.

According to the increase of the average urbanization level in middle region and the changing S_1 and S_5, Strategy 2 has obviously promoted the urbanization in middle region with S_1 shifting 0.112 to S_5. However, 9.6 % of urbanization gap still exists between middle and eastern regions.

Strategy 3. Interior Migration

Emigrate the population of middle cities which have lower urbanization than the original average urbanization level ($EX = 0.351$), and make them immigrate into middle cities which have higher urbanization than EX. Also, the

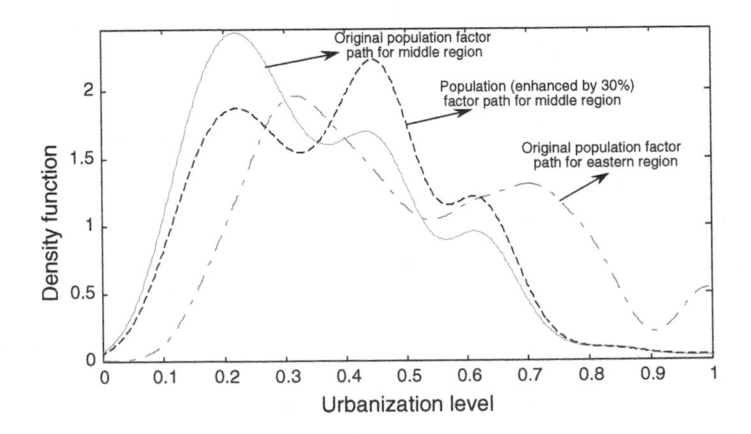

Fig. 7.2 Population factor paths for middle region with Strategy 2

Table 7.2 The effect of Strategy 2

$f(x)$	S_1	S_5	EX
$f_{\omega 0,h}(x)$	0.546	0.454	0.351
$f_{\omega 1,h}(x)$	0.465	0.535	0.377
$f_{\omega 2,h}(x)$	0.434	0.566	0.387
$f'_{\omega 0,h}(x)$	0.331	0.669	0.483

corresponding decrease and corresponding increase in each city are according to the GDP ratio.

$$p_{1i} = p_{0i} + \text{sign}(L_i - a)\frac{g_i I(L_i \geq a)}{\sum_{j=1}^{n} g_j I(L_j \geq a)} \times 0.2 \sum_{k=1}^{n} p_{0k} I(L_k < a)$$

$$p_{2i} = p_{0i} + \text{sign}(L_i - a)\frac{g_i I(L_i \geq a)}{\sum_{j=1}^{n} g_j I(L_j \geq a)} \times 0.3 \sum_{k=1}^{n} p_{0k} I(L_k < a)$$

$$\text{sign}(x) = \begin{cases} -1, & \text{if } x < 0 \\ 0, & \text{if } x = 0 \\ 1, & \text{if } x > 0 \end{cases} \tag{7.4}$$

where $a = EX$, p_{1i} and p_{2i} are a new population in middle city i through population migration by 20 and 30 %, respectively. $\omega_{1i} = p_{1i}/\sum_{j=1}^{n} p_{1j}$ and $\omega_{2i} = p_{2i}/\sum_{j=1}^{n} p_{2j}$ are new weights in middle city i when population is migrated by 20 and 30 % with project 3, respectively.

Figure 7.3 displays the comparison of new population (interior migration by 30 % with Strategy 3) factor path in middle region with the original population factor paths in middle and eastern regions.

Table 7.3 shows the effect of Strategy 3. Comparing with Strategy 1 and Strategy 2, Strategy 3 reduces the urbanization gap obviously between the middle and eastern region, since S_1 shifts 13.9 % into S_5 in the middle region.

7.3.2 Strategies of Investment Enhancement

The identifications of regional urbanization development and overall urbanization development in China validate that the investment plays a greater role in urbanization than population migration. Besides the Strategy designed through population migration, the investment also can be a measure to reduce the urbanization gap. Thus Strategy 4 and Strategy 5 are designed to reach balanced urbanization through the enhancement of investment.

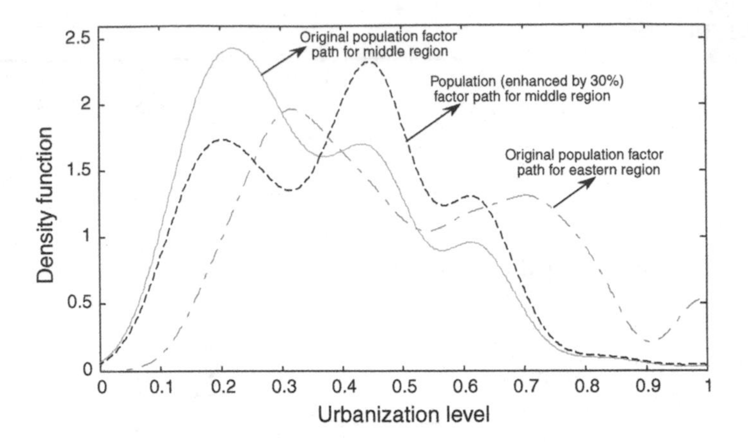

Fig. 7.3 Population factor paths in middle region with Strategy 3

Table 7.3 The effect of Strategy 3

$f(x)$	S_1	S_5	EX
$f_{\omega 0,h}(x)$	0.546	0.454	0.351
$f_{\omega 1,h}(x)$	0.454	0.546	0.380
$f_{\omega 2,h}(x)$	0.407	0.593	0.395
$f'_{\omega 0,h}(x)$	0.331	0.669	0.483

Strategy 4. Average Enhancement

Enhance the investment in middle region by 20 (or 30) percent of its original total investments. The increment in each city in middle region is distributed by the ratio of its GDP to the total GDP.

$$I_{1i} = I_{0i} + 0.2 \times \sum_{j=1}^{n} I_{0j} \times \frac{g_i}{\sum\limits_{j=1}^{n} g_j}, \quad I_{2i} = I_{0i} + 0.3 \times \sum_{j=1}^{n} I_{0j} \times \frac{g_i}{\sum\limits_{j=1}^{n} g_j} \tag{7.5}$$

where I_{1i} and I_{2i} are new investment in middle city i enhanced by 20 and 30 % with project 4, respectively.

Figure 7.4 illustrates the comparison of the new investment (enhanced by 30 % with Strategy 4) factor path in middle region, through the original investment factor paths in middle and eastern region. Table 7.4 gives the effect of Strategy 4.

$f_{2\omega 0,h}(x)$ standards for the original investment factor path in the middle region, $f_{2\omega 1,h}(x)$ the new investment (enhance by 20 %) factor path in the middle region, $f_{2\omega 2,h}(x)$ new investment (enhance by 30 %) factor path in the middle region, $f'_{2\omega 0,h}(x)$ the original investment factor path in the eastern region. S_1, S_5 and EX are exactly the same indexes in Table 7.1. The figures in Table 7.4 show that project 4 has a little effect on reducing urbanization gap between the middle and eastern region.

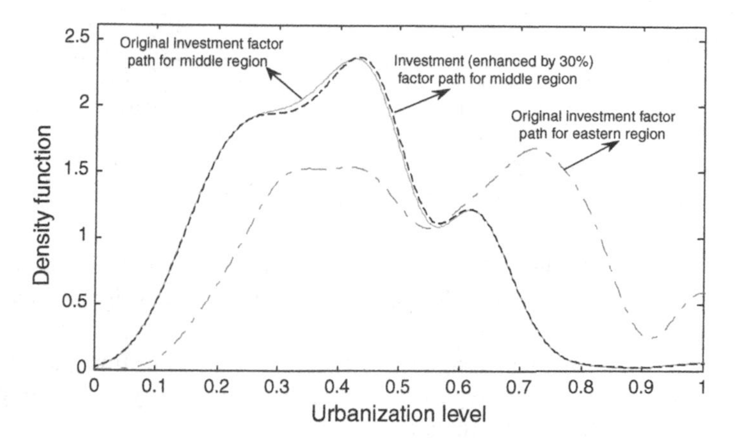

Fig. 7.4 Investment factor paths in middle region with Strategy 4

Table 7.4 The effect of Strategy 4

$f(x)$	S_1	S_5	EX
$f_{2w0,h}(x)$	0.418	0.582	0.392
$f_{2\omega1,h}(x)$	0.415	0.585	0.394
$f_{2\omega2,h}(x)$	0.413	0.587	0.394
$f'_{2\omega0,h}(x)$	0.248	0.752	0.525

Strategy 5. Selected Enhancement

Enhance by 20 (or 30) percent of its original total investments the investment in middle cities which have higher urbanization than the average urbanization level $EX = 0.392$. The increment in each city in Middle region is distributed by the ratio of its GDP to the total GDP of middle cities which have higher urbanization levels than EX.

$$I_{1i} = I_{0i} + \frac{g_i I(L_i \geq b)}{\sum\limits_{j=1}^{n} g_j I(L_j \geq b)} \times 0.2 \sum_{k=1}^{n} I_{0k}, \quad I_{2i} = I_{0i} + \frac{g_i}{\sum\limits_{j=1}^{l} g_j} \times 0.3 \sum_{k=1}^{n} I_{0k} \tag{7.6}$$

where $b = EX = 0.392$. $\omega_{1i} = I_{1i}/\sum_{j=1}^{n} I_{1j}$ and $\omega_{2i} = I_{2i}/\sum_{j=1}^{n} I_{2j}$ are the new weights for middle city i which enhance investment by 20 and 30 % with project 5.

Table 7.5 gives the effect evaluation of Strategy 5.

$f_{2\omega1,h}(x)$ stands for new investment (enhanced by 20 % with Strategy 5) factor path in middle region, $f_{2\omega2,h}(x)$ new investment (enhanced by 30 % with Strategy 5) factor path in middle region, $f_{2\omega0,h}(x)$, $f'_{2\omega0,h}(x)$ are the same indexes shown in Table 7.4, respectively. Compared with Strategy 4, Strategy 5 shows a stronger impact on reducing the urbanization gap. S_1 decreases while S_5 increases by 9 % commonly, and the urbanization gap goes down to 10.3 % between middle and eastern region.

Table 7.5 The effect evaluation of Strategy 5

$f(x)$	S_1	S_5	EX
$f_{2\omega 0,h}(x)$	0.418	0.582	0.392
$f_{2\omega 1,h}(x)$	0.353	0.647	0.414
$f_{2\omega 2,h}(x)$	0.328	0.672	0.422
$f'_{2\omega 0,h}(x)$	0.250	0.750	0.525

Figure 7.5 displays the comparison of new investment (enhanced by 30 % under Strategy 5) factor path in Middle region, through the original investment factor paths in middle and eastern regions.

Reviewing the strategies designed to reduce the urbanization gap between middle and eastern region, strategies 1–3 emphasize particularly on how migration promote urbanization while Strategies 4 and 5 stress the role of investment on urbanization. With the population migration, Strategy 3 shows an obviously effect to reduce urbanization gap and Strategy 2 is also an alternative approach to take. In the investment enhancement perspective, Strategy 5 is a better choice to reduce the urbanization gap.

7.4 Optimization of Resource Allocation in City Groups

The strategies focus on changing the investment strategy in small, medium-sized and big cities. The change of the strategy for investment will promote the urbanization development both in each individual city group and in overall China. This section presents both the studies of the individual city group and overall China through the strategies concerning the change of the investment strategy in small, medium-sized and big cities.

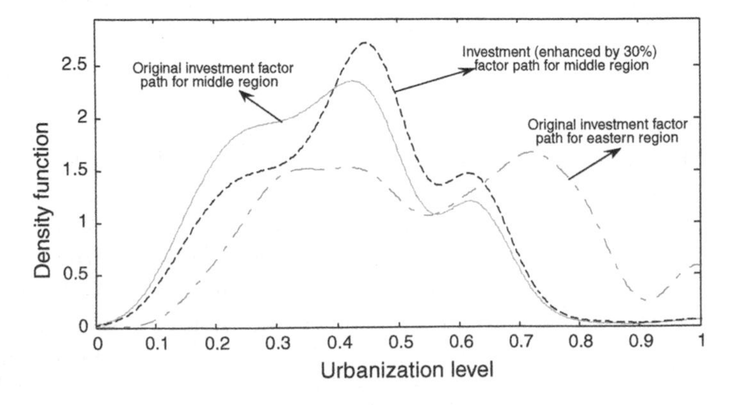

Fig. 7.5 Investment factor paths in middle region with Strategy 5

7.4.1 Individual Study

Strategy 1. Enhancement by 30 %
(a) *Extreme case.* Selected Enhancement

Make the investment in big (medium-sized or small) cities in 2003 rise by 30 %; make the increase of investment place into big (medium-sized or small) cities which have higher urbanization levels than the average level. The share of enhancement in each city is the ratio of its GDP, i.e.

$$I_{1,ki} = I_{i,k} + \frac{g_{i,k} I(L_{i,k} \geq EX_k)}{\sum_{j=1}^{n_k} g_{j,k} I(L_{j,k} \geq EX_k)} \times 0.3 \sum_{j=1}^{n_k} I_{j,k} \tag{7.7}$$

where $k = 1, 2, 3$ standards for big cities, medium-sized cities and small cities, respectively. $L_{i,k}$, $I_{i,k}$, $g_{i,k}$ and $I_{1,ki}$ mean the urbanization level, original investment, GDP and investment under project 1(A) in city i in city category k. And $EX_1 = 0.516$, $n_1 = 102$, $EX_2 = 0.323$, $n_2 = 108$, $EX_3 = 0.322$, $n_3 = 66$, respectively. $I(\cdot) = 1$ or 0, when $L_{j,k} \geq EX_k$ or the other.

(b) *Middle course.* Average Enhancement by 15 % plus Selected Enhancement by 15 %

Make the total investment in big (medium-sized or small) cities in 2003 rise by 15 %; and put the increased investment into all the big (medium-sized or small) cities according to the GDP ratio. Make another 15 % increase of the total investment in big (medium-sized or small) cities put into the big (medium-sized or small) cities with urbanization levels higher than the average level.

$$I_{2,ki} = I_{i,k} + \frac{g_{i.k}}{\sum_{j=1}^{n_k} g_{j.k}} \times 0.15 \sum_{j=1}^{n_k} I_{j,k} + \frac{g_{i,k} I(L_{i,k} \geq EX_k)}{\sum_{j=1}^{n_k} g_{j,k} I(L_{j,k} \geq EX_k)} \times 0.15 \sum_{j=1}^{n_k} I_{j,k} \tag{7.8}$$

Strategy 2. Average Enhancement by 30 % Plus Selected Enhancement by 10 %

Make the investment in big (medium-sized or small) cities in 2003 rise by 30 %; and put the increased investment into all the big (medium-sized or small) cities according to the GDP ratio. Make another 10 % increase of the total investment in big (medium-sized or small) cities put into the big (medium-sized or small) cities with urbanization levels higher than the average level.

$$I_{3,ki} = I_{i,k} + \frac{g_{i,k}}{\sum_{j=1}^{n_k} g_{j,k}} \times 0.3 \sum_{j=1}^{n_k} I_{j,k} + \frac{g_{i,k} I(L_{i,k} \geq EX_k)}{\sum_{j=1}^{n_k} g_{j,k} I(L_{j,k} \geq EX_k)} \times 0.1 \sum_{j=1}^{n_k} I_{j,k} \tag{7.9}$$

$I_{2,ki}$ and $I_{3,ki}$ are the investment under projects 1(B) and 2 in city i of city category k, respectively. Replace I_i by $I_{1,ki}$, $I_{2,ki}$, $I_{3,ki}$ in equation $\omega_i = I_i / \sum_{j=1}^{n} I_j$, gain the respective weight for city i of city category k with each project.

With the projects implementing into small cities, medium-sized cities and big cities respectively, Table 7.6 shows the average urbanization levels in each city group under different projects, which is gained by $EX = \int x f_{wh}(x)dx$. Table 7.7 presents the growth rates of urbanization level with different projects for individual study. In Table 7.7, 8.385 % = (0.38 − 0.322)/0.322, and so on.

Table 7.6 illustrates that the Strategy 1(A) achieves the greatest effect, with 5.8 % rise of the average urbanization level in small cities, 3.1 % of the one in a medium-sized city and 3.8 % of the one in a big city. However the extreme case is probably unpractical because of its unbalanced distribution of investment. The Strategy 1(B) is proposed on the basis of Strategy 1(A). It illustrates that the average urbanization level in small cities rises by 2.7, 1.6 % in medium-sized cities and the 2 % in a big city. The effect of Strategy 2 is moderate between the ones of Strategy 1(B) and Strategy 1(A).

Individually, the growth rates of urbanization levels in small cities in Table 7.7 are the highest, while the ones in big cities the lowest, Table 7.7 illustrates the strategies show the greatest effect in promoting urbanization in small cities.

On the whole, the strategies do promote the urbanization developments in small, medium-sized and big cities. Figures 7.6, 7.7, 7.8 and 7.9 also confirm the same results in Table 7.6, that is, the distributions with the original investment, the ones with Strategy 2 and the ones with Strategy 1(B) locate from left to right, reflecting a greater effect of Strategy 1(B) than the one of Strategy 2.

7.4.2 Overall Study

Getting a clear idea on how the strategies promote the individual urbanization in each city group, it is natural to come up with the idea about how the strategies work on the overall urbanization in China.

In order to promote the overall urbanization with strategies in different city groups, the study proposes to enhance the investment in small (medium-sized or big) cities according to given strategies, In order to promote the overall urbanization under projects in different city groups, the study proposes to enhance the investment in small (medium-sized or big) cities according to given projects, only

Table 7.6 Average levels with different strategies for individual study

Strategy	Small cities	Medium-sized cities	Big cities
Original	0.322	0.323	0.516
Strategy 1(A)	0.380	0.354	0.554
Strategy 1(B)	0.349	0.339	0.536
Strategy 2	0.340	0.338	0.529

Table 7.7 Growth rates of average urbanization level for individual study (%)

Strategy	Small cities	Medium-sized cities	Big cities
Strategy 1(A)	18.012	9.598	7.364
Strategy 1(B)	8.385	4.952	3.876
Strategy 2	5.590	4.644	2.519

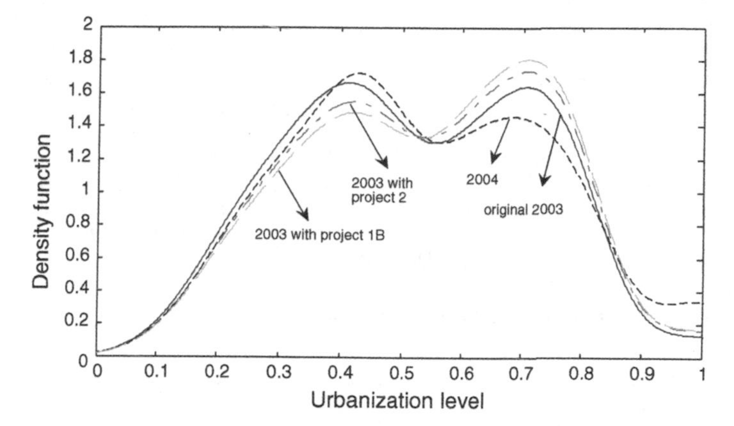

Fig. 7.6 Designed investment factor paths of urbanization in big cities

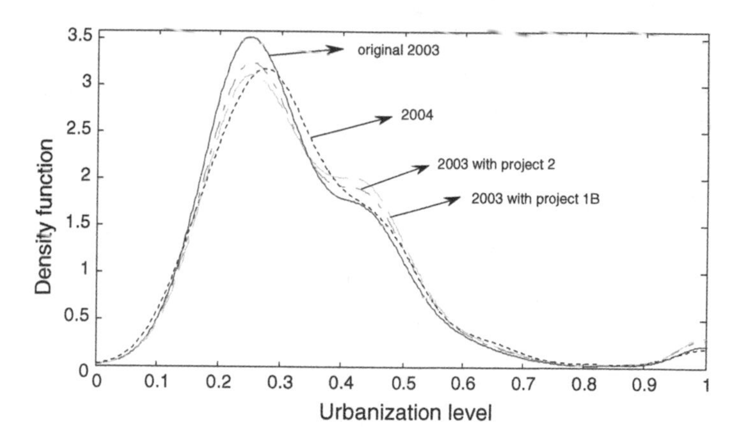

Fig. 7.7 Designed investment factor paths of urbanization in medium sized cities

by substituting EX_k by the average overall urbanization level EX, while keeping the investments in other cities the same. Then we can measures the overall urbanization with the new investments in all cities.

According to overall study, the projects are given by:

$$I_{1,ik} = I_i + \frac{g_{i.k} I(L_{i,k} \geq EX)}{\sum_{j=1}^{n_k} g_{j.k} I(L_{j,k} \geq EX)} \times 0.3 \sum_{j=1}^{n_k} I_{j,k} \tag{7.10}$$

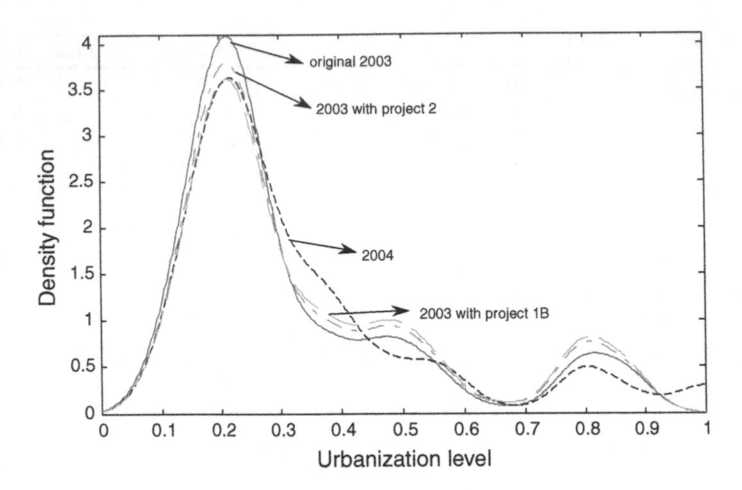

Fig. 7.8 Designed investment factor paths of urbanization in small cities

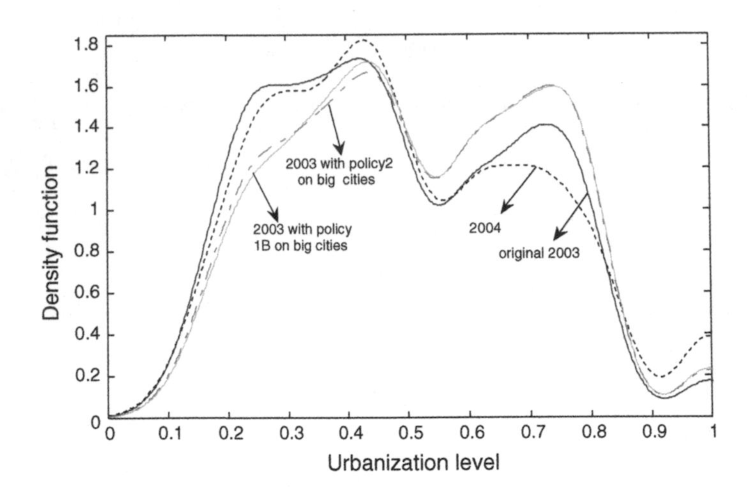

Fig. 7.9 Designed investment factor paths of overall urbanization

$$I_{2,ik} = I_i + \frac{g_{i,k}}{\sum_{j=1}^{n_k} g_{j,k}} \times 0.15 \sum_{j=1}^{n_k} I_{j,k} + \frac{g_{i.k} I(L_{i,k} \geq EX)}{\sum_{j=1}^{n_k} g_{j.k} I(L_{j,k} \geq EX)} \times 0.15 \sum_{j=1}^{n_k} I_{j,k} \tag{7.11}$$

$$I_{3,ik} = I_i + \frac{g_{i.k}}{\sum_{j=1}^{n_k} g_{j,k}} \times 0.3 \sum_{j=1}^{n_k} I_{j,k} + \frac{g_{i,k} I(L_{i,k} \geq EX)}{\sum_{j=1}^{n_k} g_{j,k} I(L_{j,k} \geq EX)} \times 0.1 \sum_{j=1}^{n_k} I_{j,k} \tag{7.12}$$

where I_i represents the original investment in city i, $i = 1, \ldots, 276$. $EX = 0.475$, $I_{1,ik}$, $I_{2,ik}$ and $I_{3,ik}$ are the new investment with project 1(A), project 1(B) and project 2 in city i in city category k, $k = 1, 2.3$ respectively.

Under the given strategies in different city groups, Table 7.8 gives the average overall urbanization level in China. Table 7.9 presents the growth rate of overall urbanization level with different strategies in different city groups. The growth rate of overall urbanization level with Strategy 1(A) in small cities is obtained by $0.632\ \% = 100\,\% \times (0.478 - 0.475)/0.475$, the figures in Table 7.9 are estimated similarly.

The average urbanization levels in Table 7.8 and the growth rate of overall urbanization levels in Table 7.9 both illustrate that the strategies in small and medium-sized cities show a little effects on overall urbanization. However, the strategies in big cities do promote the overall urbanization level obviously, with the urbanization level absolutely promoted by 3.5, 3.2 and 2.9 % and relatively increased by 7.368, 6.737 and 6.105 % with Strategy 1(A), Strategy 1(B) and Strategy 2 respectively. It is probably because the original investment in small cities or medium-sized cities shows relatively low proportion.

Figure 7.9 shows the overall urbanization distributions with the strategies in big cities. The urbanization distributions of investment factor paths show that Strategy 1(A) presents the greatest promotion and Strategy 1(B) gives a greater promotion of overall urbanization, while the Strategy 2 manifests a minor promotion of overall urbanization, which are consistent with the results given in Tables 7.8 and 7.9.

Several strategies are designed to realize the optimization of resource allocation for promoting the individual urbanization in different city groups; also the strategies are applied for overall study. Individually, the strategies do well in promoting the individual urbanization in each city group; moreover, the growth rates of urbanization levels in small cities are the highest, while the ones in big cities are the lowest, the strategies show greatest effects on promoting the urbanization in small cities. While the overall study illustrates only the strategies in big cities promote the overall urbanization level explicitly. It is probably because the original investment in small cities or medium-sized cities shows relatively low proportion.

Table 7.8 Average urbanization level with different strategy

Strategy	Small cities	Medium-sized cities	Big cities
Original	0.475	0.475	0.475
Strategy 1(A)	0.478	0.479	0.510
Strategy 1(B)	0.474	0.469	0.507
Strategy 2	0.476	0.4733	0.504

Table 7.9 Growth rate of overall urbanization level with different strategies (%)

Strategy	Small cities	Medium-sized cities	Big cities
Strategy 1(A)	0.632	0.842	7.368
Strategy 1(B)	−0.211	−1.263	6.737
Strategy 2	0.211	−0.358	6.105

7.5 Feasibility Study

The strategies do well in both reducing urbanization gap among the middle and eastern region via population migration and investment, and also in promoting the stagnant urbanization through favorable resources allocation in different city groups. However, the feasibility of strategies still bothers us a lot. The discussion in this section is provided to clarify the strategies are practical and suitable for Chinese government to implement.

On one hand, investment is introduced as an important tool for the strategies. The strategies assume the investment increase by 30 % to reduce urbanization gap among regions and the strategies make the investment increase by 30–40 % to promote urbanization among different city groups. Fortunately, the increase of investment can be achieved.

Table 7.10 describes the growth rate of total investment in fixed assets in each city group to gain the outline of the added value of total investment in fixed assets between 2003 and 2004 in each cities group. On the whole, the growth rate in each city group is around 30 %. In detail, the growth rate in small cities is 30.762 %, showing the highest; while the one in big cities is 27.142 %, showing the lowest.

Figure 7.10 illustrates the relationship between growth rate of total investment in fixed assets and urbanization level of the 276 cities at prefecture-level in 2003,

Table 7.10 Total fixed assets investment in each city group (billion Yuan)

Years	Overall cities	Small cities	Medium-sized cities	Big cities
2003	3237.437	163.9486	536.3308	2537.157
2004	4130.056	214.3824	689.8796	3225.794
Growth rate	27.572 %	30.762 %	28.629 %	27.142 %

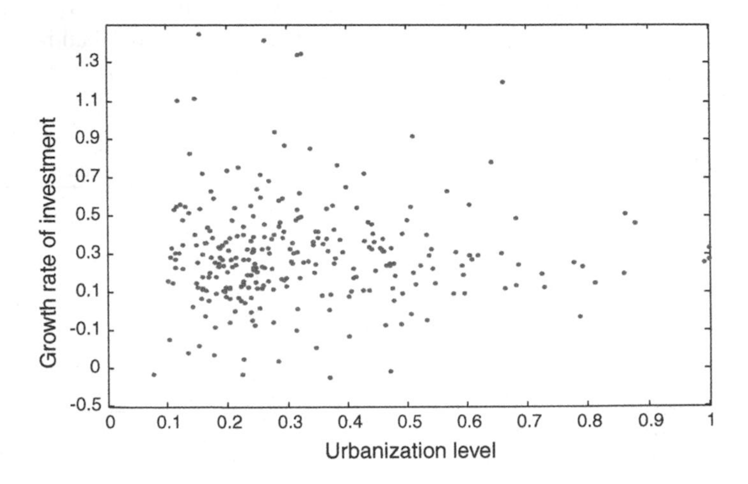

Fig. 7.10 Relationship between urbanization and growth rate of investment in 2003

showing the investment growth rates of most cities are around 30 % of their urbanization level concentrating within the interval (0, 0.4). Observing Table 7.10 and Fig. 7.10, the 30 % increase should be its natural growth rate, and the 40 % increase can be also gained by increasing the investment more aggressively.

On the other hand, the redistribution of human resources between cities and countryside and among various regions can be realized by deepening the reform of a household registration system and gradually changing the Strategy concerning change in residents. The 30 % population migration with the strategies is realizable to reduce urbanization gap in regions.

To sum up, the discussion validates the strategies is presented to reach convergence in regional urbanization development and the optimization of resource allocation to promote urbanization in different city groups.

7.6 Concluding Remarks

Synthesizing the regional differences in investment and urbanization developments between different regions and city groups will be more comprehensive in measuring the changes in production efficiency under sustainable growth background. This chapter aims to pursue sustainable production efficiency in china with optimal resource allocation and balanced urbanization development via the transplantation strategies with both investment and population agglomeration.

To realize the balanced development via reducing the regional urbanization gap between middle and eastern regions, three strategies emphasizing on population migration and two strategies concerning investment enhancement are designed. On the whole, the study results in that Strategy 3 for population and Strategy 5 for investment will be two promising strategies to narrow the urbanization gap between middle and eastern regions. Strategy 3 carries out an interior migration of population by 30 % and Strategy 5 takes a selected enhancement of investment by 30 %. Strategies 3 and 5 narrow the regional urbanization gap by 4.4 % [= (48.3 − 35.1 %) − (48.3 − 39.5 %)] and 3 % respectively between middle and eastern regions.

To realize the optimization of resource allocation through promoting urbanization in different city groups, the study proposes two investment strategies for promoting the urbanization both for each city group individually and for the overall China.

Individually, the strategies do well in promoting the individual urbanization in each city group; moreover, the growth rates of urbanization levels in small cities are the highest, while the ones in big cities the lowest.

While the study of overall China illustrates only the strategies in big cities promote the overall urbanization level obviously. The overall urbanization level is absolutely promoted by 3.5, 3.2 and 2.9 % and relatively increased by 7.368, 6.737 and 6.105 % with Strategy 1(A), Strategy 1(B) and Strategy 2, respectively. Strategy 1(B) makes an additional 10 % rise investment put into big cities with urbanization level than the overall average urbanization level, and Strategy 2 only changes 50 % of the original added value of investment in big cities.

The feasibility of the strategies is settled by deepening the reform of a household registration system and the natural investment growth rates of most cities being around 30 %.

References

Catin, M., Luo, X. , Huffel C. V. (2005). Openness, industrialization, and geographic concentration of activities in China. Policy research working papers, No.3706,1–23.

Chen, Q. (2006). Crisis and opportunities of the urbanization in China. *City Planning Review, 1*, 34–39.

Lu, M., & Chen, Z. (2004). The economic policy of urbanized, urban tendency and income difference between city and countryside. *Economic Research Journal, 6*, 41–49.

Li, S., Jiang, S., & Hu, Y.(2005). An quantitative analysis on the demand of infrastructure construction Investment in the process of urbanization. Urban Studies, 12(4), 232–254.

Li, Z. (2007). The regional differences and factors analysis on urbanization level in China. Yunnan Geographic Environment Research, 19(1), 85–91.

Liu, G., Yang, J., & Deng, X. (2006). Study on Economic Factors Affecting Urbanization in China. Urban Studies. 13(5), 9–12.

Kim, T., & Lee, S. (2005). Kernel density estimator for strong mixing processes. *Journal of Statistical Planning and Inference, 133*(2), 273–284.

Xu, B., & Cai, G. (2007). Maximal inequality and complete convergences of non-identically distributed negatively associated sequences. *Applied Mathematics A Journal of Chinese Universities, Ser. B, 22*(3), 316–324.

Zheng, L. , Huang, Q., Lu, T., & Zhou, W. (2007). The process and problems of industrialization and urbanization in china. China Industry Economy, 40(1), 6–30.

Zhu, S. (2006). Analysis on relativity between urbanization and developing difference of regional economy, Value Engineering, 7, 11–13.

Chapter 8
Concluding Remarks

The book "Changes in Production Efficiency in China: Identification and Measuring" provides decision support on investment and population agglomeration strategies with unique technique of *Path Converged Design*. The contribution of this book is to identify and measure how investment and population agglomeration changes in production efficiency in sustainable growth from aspects of allocation efficiency, scale efficiency, technical efficiency, and sustainable efficiency. The innovations, results and applications of the book are summarized as below.

8.1 Innovations

Since Farrell's pioneering research of production efficiency, various researches have been made to compute and analyze production efficiency from economics and statistics perspectives. However, no application-based studies were found in previous literatures to identify and measure the changes in production efficiency from perspective of management engineering.

To realize production efficiency, the first difficulty is that the production frontier surface, the criterion in existing measurement of production efficiency, cannot be observed because it is usually an ideal concept rather than a real object. The second difficulty is that it is hard to control the variables and isolate the relations to identify an unknown underlying production function because natural variables usually cannot be reproduced in laboratory. Hence the difficulties often result in several contradict conclusions due to uncontrollable variables.

Contradict conclusions concerning the analyses of production efficiency are permissive and significant in economics. However, decision maker is impossible to reach any decision facing different identified conclusions. An original path identification technique is presented in this book. It is the first study that attempts to realize production efficiency from engineering and statistics perspectives rather than from economics and mathematics ones. First, it presents an observable benchmark as the criterion of the production efficiency to replace the unobservable production frontier surface. Second, path-converged design is designed to select a

B. Xu et al., *Changes in Production Efficiency in China*,
DOI: 10.1007/978-1-4614-7720-4_8,

controllable variable as a path of identification and to ignore uncontrollable natural variables.

Different from previous approach, two innovations are presented as follows.

- Criterion of efficiency: Benchmark model

Theoretically, production efficiency will include all points along the production possibility frontier, but it is difficult to measure in practice because the actual production can be observed easily while the production frontier surface can't.

With efficiency-oriented framework rather than output-oriented or input-oriented one, the observable benchmark model is presented to replace the production frontier surface, which can achieve criterion of production efficiency in cities or regions through the distribution of production function rather than the parameters.

- Technique of identification: Path model

The path identification, which is established through the design of path model converging to benchmark model, can successfully be applied to implement optimal strategy of investment and migration to measure the changes in production efficiency by controllable inputs in cities or regions in China.

8.2 Results

The book contributes to management engineering most significantly in identifying and measuring how investment and population agglomeration changes in production efficiency in sustainable growth from aspects of allocation efficiency, scale efficiency, technical efficiency, and regional equity with path converged techniques, which provides decision support in urban planning. The main realized results are presented as below.

- *Allocation efficiency with investment and population agglomeration*

The implementations of additional investment of PE successfully reduce their allocation inefficiency strengths to 2 % by additionally investing PE of 101.76 billion Yuan for overall China in 2006, which are 7.380 % of their original total amount (Chap. 2).

Decision-making for regional population migration performs well in eliminating allocation inefficiency of urbanization development. By emigrating about 14, 10, and 14 % of the regional population from cities at low urbanization levels to cities at higher urbanization levels, allocation inefficiency strengths between benchmark and regional population distributions shrink to 5.800, 4.100 and 5.600 % from 14.64, 9.850 and 13.97 % for small, medium, and large cities, respectively (Chap. 3).

- *Scale efficiency with investment*

The identification reveals that regional production efficiency discrepancy is mainly due to capital elasticity with FDI. FDI path witnesses crowd-in effect on capital after 1996 in Eastern region while crowd-out effect on capital in both Middle and Western regions during almost all the time. Possible decision makings for narrowing regional discrepancy are designed with environment FDI path identification. Implement of urbanization environment with FDI path realizes the crowd-in effect on capital in both Middle and Western regions (Chap. 4).

- *Technical efficiency with investment*

Technical progress in Eastern region is significant with FDI path because the declining trend of technical level is reversed after 1994. Unfortunately, the technical progress does not appear in either Middle or Western region. The TFP growth with FDI path is mainly attributed to technical progress rather than efficiency improvement in both Eastern and Middle regions. Furthermore, realizations of technical progress are obtained with path-converged design in Middle and Western regions (Chap. 5).

Identification illustrates FDI path produces increasing technical efficiency and avoids the diminishing marginal returns in Eastern region, while both cases fail in Middle and Western regions. Decisions are realized through transplantation of Eastern technical efficiency into Middle and Western regions not only succeeds in producing output growth, but also keeps the production efficiency from decreasing. The transplantation of scale efficiency only succeeds in producing output growth while fails in reversing the declining trend of production efficiency (Chap. 6).

- *Sustainable efficiency with urbanization*

To pursue sustainable production efficiency in China, the transplantation strategies with both investment and population agglomeration are presented to reach optimal resource allocation and balanced urbanization development. An interior migration of population by 30 % and a selected enhancement of investment by 30 % will be two promising strategies to reduce the regional urbanization gap between middle and eastern regions and hence to realize balanced and sustainable growth. An additional 10 % rise investment put into big cities with urbanization level than the overall average urbanization level and only changes 50 % of the original added value of investment in big cities will be two promising strategies in big cities promote the overall urbanization level for sustainable efficiency (Chap. 7).

8.3 Convergence Speed of Path Identification

The original technique of path identification can also be applied in extensive areas, such as database design in information engineering, risk control in finance engineering and catastrophic safety warning in decision engineering, which will bring about many interesting researches.

The research in this section presents the investigation of return rates in stock market with the convergence speed of path identification technique. Data of daily turnover and closing prices on April 21–22, 2009 are employed in this study from http://www.stockstar.com/. The data covers 164 stocks that belong to SSE-180 Index in China.

Using path converged design in Chap. 2, let $X_i = \ln p_i(t) - \ln p_i(t-1)$, $i = 1, \ldots, n$, be the return rate of stock i, $y_i = X_i/T_i$, T_j refers to the turnover (price $\times$ volume) of stock i, $\omega_i = y_i/\sum_{j=1}^{n} y_j$ is taken as the **turnover path** with efficiency-oriented production for stock i, $i = 1, 2, 3, \ldots, n. n = 146$. The benchmark model of return rate is presented by kernel density estimation:

$$f_n(x, X) = \frac{1}{n}\sum_{i=1}^{n} h^{-1}K\left(\frac{x - X_i}{h}\right) \quad (8.1)$$

where $K(\cdot)$ stands for kernel function and h the bandwidth. The turnover path model of return rate is given by:

$$f_n(x, X, \omega) = \sum_{i=1}^{n} \omega_i h^{-1}K\left(\frac{x - X_i}{h}\right) \quad (8.2)$$

Figure 8.1 presents the turnover path identification of return rate for SSE-180 Index on April 21st. The distribution curve presents its first peak at −0.3, which indicate lots of turnover takes place at return rates interval (−0.5, −0.3). If the Efficient Market Hypothesis (EMH) holds true, a lot of turnover will converge to the return rate around zero, which is the second peak of the distribution. In this sense, the situation in Fig. 8.1 implies the market might need adjustment.

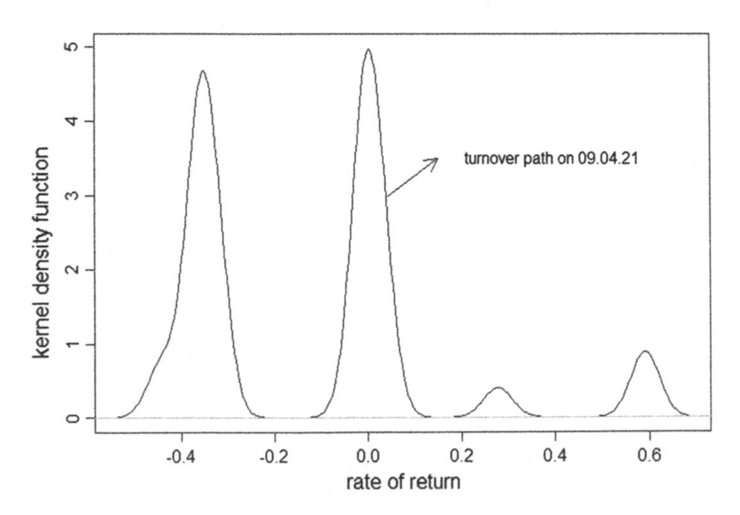

Fig. 8.1 Turnover path identification of return rate for SSE-180 Index

How to forecast and realize the adjustment of turnover is of crucial importance to investors. Figure 8.2 illustrates the turnover path identification of return rate for SSE-180 Index centers more symmetrically to the return rate of zero than the benchmark identification distribution. It points the adjustment orientation for benchmark identification. The reallocation of turnover is provided with the idea in Chap. 2.

Figure 8.3 gives the reallocation turnover path identification performs well in forecasting the market trend on forthcoming day while fails to adjust adequately. Since the reallocation turnover path identification curve on April 21st shifts right and shares similar changing trend to real turnover path identification curve on April 22nd.

To validate application of path identification, Fig. 8.4 lists both the realization of convergence with simulation and the realization of forecast of return rate for SSE-180 Index on April 23rd. First and foremost, the simulation is implemented first to sample 360 random observations from the turnover path distribution on April 22nd, and second to construct new benchmark and new turnover path identification of return rate. The simulation successfully realizes the convergence of path model to benchmark model because the two curves are close to each other. Furthermore, the realization of convergence with simulation validates the possibility of realization of forecast in finite sample. The reallocation turnover path identification curve on April 21st is very similar to the actual turnover path identification of return rate on April 23rd, which realizes the forecast of return rate on forthcoming days.

Although the benchmark model and path model in this book have presented promising results in the path identification towards identifying and measuring the changes in production efficiency in China, two specific points deserve further investigation.

Firstly, the book mainly explores the one-path identification, while the multi-path identification may be more interesting and deserves more exploration.

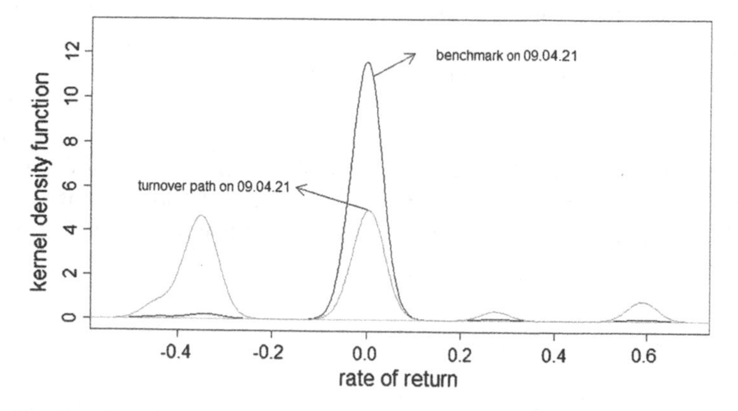

Fig. 8.2 Benchmark and turnover path identifications of return rate for SSE-180 Index

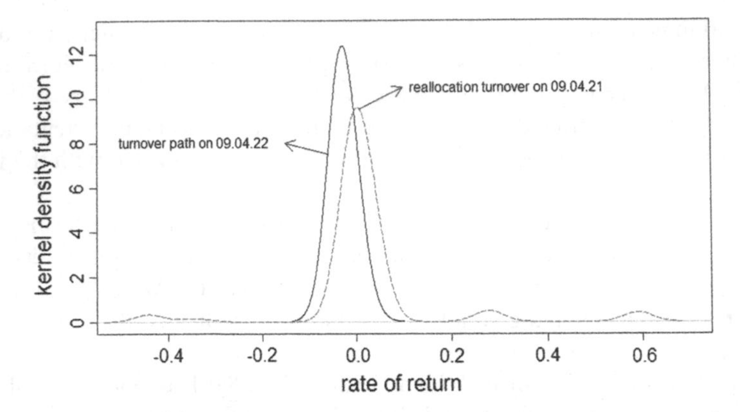

Fig. 8.3 Reallocated turnover path identification of return rate for SEE-180 Index

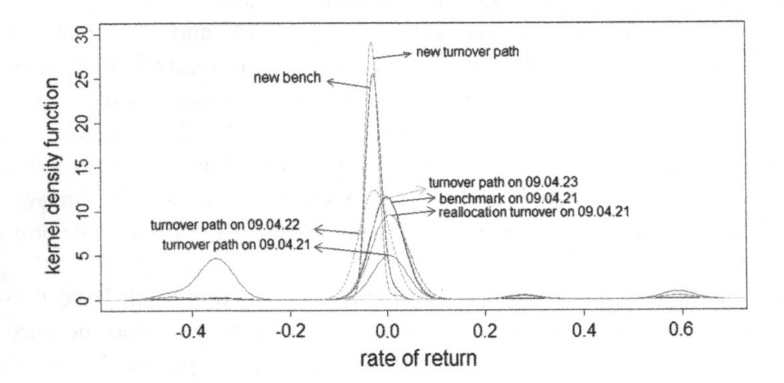

Fig. 8.4 Realization of convergence and forecast of return rate for SSE-180 Index

However, multi-path identification will be much more complex involving the interrelationships among different paths.

Secondly, the identification of the underlying production function with the path-converged design can be implemented in many applications. However, the identification is obtained dependent on the nonparametric approach. It is well accepted that some steps are still indispensable in the nonparametric techniques, such as the selection of bandwidth, which is difficult for given data. Statistical theories of large sample are mainly applied to select the optimal bandwidth by R software in this book, which may be misleading for finite observations. In addition, the statistical features of the nonparametric estimations, such as the one for confidence intervals, need further discussion.

References

Aigner, D. J., Lovell, C. A. K., & Schmidt, P. (1977). Formulation and estimation of stochastic frontier production function models. *Journal of Econometrics,6*(1), 21–37.

Dierk, H., Stephan, K., & Felicitas, N. D. (2007). In search of FDI-led growth in developing countries. *Economic Modelling,25*(5), 793–810.

Economic Growth Frontier Subject Team (2004). The supply effects of fiscal policy and economic development. *Economic Research Journal9*, 4–17

Engle, R. F., & Granger, C. W. J. (1987). Cointegration and error correction: Representation, estimation, and testing. *Econometrica,55*, 251–276.

Espino, D., & Toribio, R. (2004). Determination of technical efficiency of fisheries by stochastic frontier models: a case on the Gulf of Cadiz (Spain). *ICES Journal of Marine Science,61*(3), 416–421.

Griffin, J. E., & Steel, M. F. J. (2008). Flexible mixture modelling of stochastic frontiers. *Journal of Productivity Analysis,29*, 33–50.

James, B. A. (2008). What are the mechanisms linking financial development and economic growth in Malaysia? *Economic Modelling,25*(1), 38–53.

Koop, G., Osiewalski, J., & Steel, M. (2000). A stochastic frontier analysis of output level and growth in poland and western economies. *Economics of Planning,33*, 185–202.

Kuo, C.-C., & Yang, C.-H. (2008). Knowledge capital and spillover on regional economic growth: Evidence from China. *China Economic Review,19*(4), 594–604.

Lu, W., & Lo, S. (2007). A closer look at the economic-environmental disparities for regional development in China. *European Journal of Operational Research,183*, 882–894.

MacKinnon, J. G. (1996). Numerical distribution functions for unit root and cointegration tests. *Journal of Applied Econometrics,11*, 601–618.

Nemoto, J., & Goto, M. (2005). Productivity, efficiency, scale economies and technical change: A new decomposition analysis of TFP applied to the Japanese prefectures. *Journal of the Japanese and International Economies,19*, 617–634.

Wacker, J., Yang, C., & Sheu, C. (2006). Productivity of production labor, non-production labor, and capital: An international study. *International Journal of. Production Economics,103*(2), 863–872.

Wang, C. (2007). Decomposing energy productivity change: A distance function approach. *Energy Economics,32*, 1326–1333.

Wei, Q., Yan, H., & Xiong, L. (2008). A bi-objective generalized data envelopment analysis model and point-to-set mapping projection. *European Journal of Operational Research,190*, 855–876.

Xin, L., & Deng, S. (2007). On crowding-in effect and crowding-out effect of FDI to private investment. *Modern Finance and Economics,27*, 9–14.

Xu, B., and Watada, J. (2008), The Path Identification of FDI and Financial Expenditure in China, International Symposium on Management Engineering 2008 (ISME2008), Kitakyushu, Japan

B. Xu et al., *Changes in Production Efficiency in China*,
DOI: 10.1007/978-1-4614-7720-4,

Xu, B., & Cai, G. (2007). Maximal inequality and complete convergences of non-identically distributed negatively associated sequences. *A Journal of Chinese Universities,22*, 316–324. Ser. B.

Xu, B. and Watada, J. (2009). Decision-making for the optimal strategy of population agglomeration in urban planning with path-converged design. In C. P. Lim, L. C. Jain (Ed.), *New directions in decision support systems: Methodologies and applications.* Germany: Springer.

Xu, B., & Watada, J. (2008). Liquidity impact on sector returns of stock market: Evidence of China. *Global Journal of Finance and Banking Issues,2*(2), 56–69.

Printed in the United States
By Bookmasters